THE SECRET LIFE OF DUST

*From the Cosmos to the Kitchen Counter,
the Big Consequences of Little Things*

Hannah Holmes

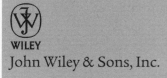

WILEY

John Wiley & Sons, Inc.

The poem "Dust" appearing on page v copyright © 2002 by Thomas Carper. Printed with the author's permission.

Published by John Wiley & Sons, Inc., Hoboken, New Jersey
Published simultaneously in Canada

For general information about our other products and services, please contact our Customer Care Department within the United States at (800) 762-2974, outside the United States at (317) 572-3993 or fax (317) 572-4002.

Wiley also publishes its books in a variety of electronic formats. Some content that appears in print may not be available in electronic books. For additional information about Wiley products, visit our website at www.wiley.com.

Library of Congress Cataloging-in-Publication Data:
Holmes, Hannah
    The secret life of dust : from the cosmos to the kitchen counter, the big consequences of little things / Hannah Holmes.
        p.   cm.
    Includes bibliographical references and index.
    ISBN 0-471-37743-0 (cloth : acid-free paper)
    ISBN 0-471-42635-0 (paper)
    1. Science—Miscellanea.   2. Dust—Miscellanea.   I. Title.
Q173.H733   2001
551.51'12—dc21   2001022368

Printed in the United States of America

10 9 8 7 6 5 4 3 2 1

To my big, fat muse, P. Earth

DUST OR ON FIRST LOOKING INTO HANNAH HOLMES'S *THE SECRET LIFE OF DUST*

There are no golden harps to strum,
Ambrosial wines and quiches,
Elysian fields, or kingdoms come,
Or otherworldly geishas:

Only the comfort of a slow
Particulate dispersal,
For which our timely death, we know,
Is merely a rehearsal.

Yet when millennia erase
The ground that has o'er-hovered us,
A certain future, by the grace
Of wind, will have discovered us.

For out into expanding void
Beyond our fading star,
We'll leave the planet we enjoyed
Without an "au revoir"

To be, at last, eternal dust—
A notion not alluring,
But, in our cosmic bang-to-bust,
Still strangely reassuring.

—Thomas Carper

CONTENTS

ACKNOWLEDGMENTS

My profound gratitude goes to the hundreds of scientists who spent time talking with me, digging up obscure papers, and pointing me in fertile directions. Without those who gave so freely of both their knowledge and their enthusiasm for the subject, this book would not exist. Some of these dust scholars dedicated many precious hours to my education and to improving the manuscript, for which I am even more deeply indebted. They bear no blame for any inaccuracies that may remain. They are:

Chapter 2: David Leckrone, stargazer extraordinaire and a dear friend. Also Stanley Woosley, Neal Evans, Henry Throop, David Leisawitz, Max Bernstein, and Kim Sepulvur. Chapter 3: Don Brownlee, Susan Taylor, Ken Farley, Mike Zolensky, and Larry Nittler. Chapter 4: David Loope, David Fastovsky, and Marc Hendrix. Chapter 5: Estelle Levetin, Barry Huebert, David Pyle, and Hayley Duffell. Chapter 6: Dan Jaffe, Steve Warren, Douglas Westphal, and Joe Prospero. Chapter 7: Pierre Biscaye, Dean Hegg, Tamara Ledley, and David Rind. Chapter 8: Gene Shinn, Daniel Muhs, Richard Schlesinger and Morton Lippmann, Garriet Smith, Mary Silver, John Priscu, David Miller, and Charles Main. Chapter 9: Robert Castellan, Susanna Von Essen, Eileen Schneider, and Albert Heber. Chapter 10: Paul Lioy, Andy Liu, Lance Wallace, Eileen Abt, Bernard Harlow, Frank Vigil, Jens Ponikau, John Roberts, and Agile Redmon. Chapter 11: Lee Anne Willson, Fred Adams, and Ken Caldeira.

Special thanks to David Vardeman at the Portland Library of the University of Southern Maine, for reeling in some of the most esoteric journals on the planet. And thanks to all the staff there, for delivering the printed world to my doorstep.

Finally, thanks to my parents, whose kids grew up with a microscope on the kitchen table. Thanks to my friend and agent, Karen Nazor, for the kick in the pants. Dad and Kirsten, thanks for reading! And for cheerfully braving unreadable chapters, for sharp-eyed guidance, and for ensuring that Mr. Mookie Moe got his beach time, thank you to the Big Fish, Claude V. Z. Morgan.

INTRODUCTION

What got me started on dust?

The little subject suggested itself rather forcefully. A few years ago I was on assignment in Mongolia's Gobi Desert, writing about a dinosaur expedition. The pink-orange clouds of dust that billowed over the desert floor were impossible to ignore. The whirling specks invaded my eyes and nose. They infiltrated the pages of my books. They invaded the depths of my sleeping bag.

I had thought this rambling dust was just a local phenomenon. So I was immediately intrigued when expedition geologist David Loope told me that a thin veil of dust flows high across the sky, enveloping the entire planet. As we stood by a sandstone cliff, squinting against the ever-swirling grit, Loope explained how dust helped to create the Gobi's fabulous fossils. Raindrops form on these high-flying dusts, he said. The falling rain drags down the dust, and that dust works a dark magic inside a sand dune.

Consider the larger implications of this: Worldwide, how many raindrops fall from the sky on any given day? And each raindrop contains a piece of dust. So how many specks must there be in the sky? Where do they all come from?

Another of our expedition mates warned me that I'd be digging Gobi dust out of my ears for six months after I left the desert. But some of the dust seems to have penetrated even more deeply. Back at home I discovered I had dust on the brain. When I look up at the sky, I search for a glimmer of that ubiquitous veil. When a raindrop pelts my arm, I stare at the splattered water, wondering what sort of speck brought this drop together. When I wipe my computer screen, I peer through a magnifying glass at the sparkly, fuzzy stuff caught among the sharp ridges of my fingerprint. Too small to distinguish are the individual fragments of a disintegrating world: the skin flakes, rock flecks, tree bark, bicycle paint, lampshade fibers, ant legs, sweater wool, brick shards, tire rubber, hamburger soot, and bacteria. The world is in a constant state of disintegration.

This invisible dust isn't as harmless as it may appear. It can be a heartless little brute. In -ologies ranging from climatology to immunology, scientists are now calling dust onto the carpet. It is a central suspect in the mystery of how the planet's climate is shifting. Billions of tons of it take to the sky each year, and this surely alters the behavior of the Earth's atmosphere. And dust is taking new heat for killing lots of people—not just miners, sandblasters, and asbestos workers, but thousands, maybe *millions,* of ordinary people who simply live and breathe in dusty air. Although our bodies evolved to screen out most natural dusts, it seems that our lungs are vulnerable to the smaller, industrial-size specks. Dust's relationship to asthma is another topic now coming to a boil. Traditionally, scientists thought the asthma epidemic might be caused by various house dusts. But new, jaw-dropping evidence suggests that asthma may be caused by *too little* house dust.

By necessity dust scholars are a creative bunch. Scientists who study elephants are spared much of the difficulty of locating a specimen. But a dust scientist must often invent a device simply to *acquire* the minuscule object of his or her curiosity. One woman created an underwater vacuum cleaner to collect her space dust from the bottom of a well. A fellow who studies the dust of the last ice age isolates his tiny samples of dust from glacial ice cores. And catching dust is only half the battle: both the handling and the analysis of dust are complicated by its dainty girth. The latter scientist rounds up his flighty grains with cling-wrapped fingers.

Since the day I stood in the Gobi Desert and contemplated the population of dust in the sky, I've come to see the air as a medium and dust as the message. Dust delivers the world news: the Rocky Mountains are eroding, and a volcano is erupting in the Philippines. It carries the local headlines, too: the neighborhood coffee roaster is burning the beans, and traffic is heavy on the turnpike. And it brings us the social pages, the news about human activity, for we are dusty creatures.

One purpose of this book is to help readers learn to decipher some of the messages that drift in the air. Our planet sometimes seems too enormous to really comprehend. But perhaps tuning in to the news bulletins issued by some of the planet's smallest reporters can give us a better sense of how things are going for the whole.

Second, I'd be honored if I were able to introduce the reader to his own, personal dusts. Never mind that each of us is constantly enveloped in a haze of our own skin flakes and disintegrating clothing. In addition to that cloud, each match we strike, each light-switch we flick, and each mile we drive causes more dust to rise into the air. Taken in global quantities, our personal puffs of dust have planet-size consequences.

When the fragmenting skin of the Earth rises, both at nature's urging and our own, it changes the weather, and even the climate. When it settles, this dust alters the seas and the soils and the delicate linings of our own lungs. In tiny things there is huge magic and colossal mayhem.

A few notes on jargon:

- Temperatures will be given in Fahrenheit.

- The subject of size is covered in Chapter I. But for ease of reference, here is a sampling of small things:

One inch:	25,000 microns
A period:	In this font 300 microns
Sand:	63 microns and larger
Dust:	63 microns and smaller *
Human hair:	100 microns **
Pollen:	10–100 microns
Cement dust:	3–100 microns
Fungal spores:	1–5 microns
Bacteria:	0.2–15 microns
Fresh stardust:	0.1 microns
Various smokes:	0.01–1 microns
Tobacco smoke:	0.01–.5 microns

- "Sulfur beads." Many scientists have cautioned me against including what I call sulfur beads in my broad definition of dust. The objection is understandable: When sulfur gases from a coal fire or an erupting volcano condense into little clumps in the sky, the clumps are often liquid, because they quickly draw water from the atmosphere around them. But in dry air, sulfur does indeed form dry particles. In fact, a single sulfur particle can gather or lose water as it travels through the sky, turning from liquid to solid and back again. To scientists, these changeable little items are "aerosols." But since this is not a technical book, I can't see the harm in including them in the dust family.

---

* Technically, what I call dust a geologist would divide into "silt" and "clay." Some geologists draw the sand/silt line at 63 microns, others at 60 or even 50. Most agree that "clay" is smaller than 4 microns.
** This varies widely from person to person.

1

THE WORLD IN A GRAIN OF DUST

Picture a juice glass sitting on a porch railing in the sunshine. It may look empty, but churning inside that glass are twenty-five thousand microscopic pieces of dust—at *least*. And these dusts are a little bit of everything on Earth. One minute they're tiny crumbs chipped off Saharan sand and invisible shreds of camel hair. Then the wind shifts, and you are surrounded by spores of forest fungi and fragments of desiccated violets. A bus stops nearby to take on passengers, and flakes of human skin mixed with minuscule specks of black soot momentarily dominate.

Every time you inhale, thousands upon thousands of motes swirl into your body. Some lodge in the maze of your nose. Some stick to your throat. Others find sanctuary deep in your lungs. By the time you have read this far, you may have inhaled 150,000 of these worldly specks—if you live in one of the cleanest corners of the planet. If you live in a grubbier region, you've probably just inhaled more than a million.

Although these dusts have been waved aside for most of human history, in this book we'll see that dust is terrifically consequential. Some dusts menace the planet and its living residents. Some are beneficial to people, plants, and animals. Many are merely fascinating. All are going under the microscope. And the secret lives of dust are being revealed.

One of the most impressive revelations is how much dust surrounds us—the sheer tonnage of stuff rising off the face of the Earth. Because these specks are so small and shifty, the estimates are still rough. Nonetheless, irrefutably *huge* amounts of small things take to the wind each year.

Between 1 and 3 billion tons of desert dust fly up into the sky annually. One billion tons would fill 14 million boxcars, in a train that would wrap six times around the Earth's equator.

Three and a half billion tons of salt flecks rise off the oceans.

Trees and other plants exhale a billion tons of organic chemicals into the wind, perhaps one-third of which condenses into tiny, sailing beads.

Plankton, volcanoes, and swamps leak 20 to 30 million tons of sulfur compounds, about half of which forms little airborne specks.

Burning trees and grasses throw up 6 million tons of black soot.

The world's glaciers slowly grind their host mountains into dust that takes to the wind—but in what quantities? No one knows.

Likewise, how many glassy bits of volcanic ash are blasted into the ether?

And the dusts of life—flying fungi, viruses, diatoms, bacteria, pollen, fibers of rotting leaves, eyes of flies and legs of spiders, scales from the wings of butterflies, hair fragments from polar bears, skin flakes from elephants—how many tons of these roam the atmosphere?

About 4 million years ago our ancestors began to augment the dusty exhalations of nature. At first we supplemented the soot, as we mastered the mesmerizing tool of fire. Then, when we learned about the miracle of metals, our smokes grew richer with microscopic beads of hot bronze, iron, copper, gold, and silver. The advent of spinning and weaving produced invisible fragments of animal and plant fibers, which the wind lifted out of our encampments. Finally, with the industrial revolution, our dust output shifted into high gear.

Ninety to 100 million tons of sulfur now rise annually from the world's fossil-fuel burners—mainly coal-fired power plants, but also oil-fired plants and diesel engines. Every natural sulfur bead in the sky is now accompanied by between three and five human-made beads. And the Earth hosts more fuel burners every day.

More than 100 million tons of nitrogen oxides, which like sulfur gas are prone to form dusty particles in the sky, flow upward from our farms, and automobiles and other fuel-burning inventions.

Eight million tons of black soot in the sky are attributable not to burning trees and grasses but to the conflagration of fossil fuels—especially coal. Even of the 6 million tons of soot that rain upward from tree-and-grass fires, most can be traced to the human hand.

Whether the skies carry 1 billion or 3 billion tons of desert dust, fully half may be our responsibility. Our agriculture and other assaults on the landscape may have doubled the amount of desert dust naturally present in the air.

And the miscellaneous dusts of the twentieth century—nerve-racking mercury and stupefying lead; carcinogens from dioxin to polychlorinated biphenyls (PCBs); the radioactive dusts of nuclear disasters, pesticides, asbestos, and poisonous smokes—how many tons of these roam the skies each year? That is unknown.

. . .

If the quantities of dust are hard to gauge, dust scholars have an easier time pinning a *size* on various dusts. Generally, the dusts that whirl around us are so small that gravity has to fight to get control of them. Forces on the surface of a piece of dust—static electricity, even the interaction of one atom with another—can overpower the call of gravity. Dust can perch on the ceiling as easily as on the tabletop.

Scientists measure dust in microns, or twenty-five-thousandths of an inch. Consider the hair on your arm. A single hair might be 100 microns wide. Now imagine taking up scissors and snipping off a section 100 microns *long*. That tiny snippet, visible only if you know where to look for it, is too big to be dust. From a scientist's perspective, that snippet falls in the family of sand.

The very biggest grains of dust are, technically, only two-thirds as wide as a hair. These fat dusts are usually the work of nature. The diameter of pollen grains, for instance, ranges from a full hair's width to a tenth of a hair's width. If you sift a handful of sand from the beach or the desert, the faint powder that sticks to your palm will be a range of sizes, with lots of grains in the fatter category. The flakes of dead skin that float out through the weave of your shirt to form an invisible halo around you are rectangles one-tenth of a hair wide and two-tenths of a hair long. Many of the salt flecks that blow off the oceans are upward of 5 microns wide. And those are still some of the larger dusts.

Health scientists fret more about small dusts than large ones. That's because the human body has evolved to bar the entrance of nature's big creations. Nearly all pollens, for example, are so big that they get hung up inside the nose—as people with allergies are well aware. But small dusts can slide right past the traps inside your head and sail deep into your delicate lungs.

Until recently, scientists drew the line between safe and dangerous dusts at ten microns—one-tenth of a hair's width. But as dust investigators peered more closely at their little subjects, they decided to move the line. Medical research now shows that dusts less than one-quarter that big—a twenty-fifth of a hair—cause the most disease and death. Even as scientists rewrite dust limits to protect our lungs, they're still struggling to understand *how* tiny dusts kill.

So which dusts fall on the small side of the line? A few natural dusts make the cut: bacteria and fungal spores are usually well under 10 microns. But industrial dusts are the dominant force in the "teensy" category. Pesticide dusts are often between half a micron and 10 microns wide. The very biggest particles in a puff of tobacco smoke are less than half a micron wide—that's one two-hundredth of a hair. The smallest particles in automobile exhaust are a hundredth of a micron—one ten-thousandth of a hair. This is also the realm of tiny particles that form when pollution gases condense into beads in the air. Viruses and big molecules are about the same size. You can begin to imagine how 25,000 of these tiny motes could roam a juice glass unnoticed.

For all the murder and mischief we'll see it commit in this book, dust is nonetheless indispensable. The Sun we circle was created inside a giant womb of protective space dust. Some of that same dust—tiny specks the size of cigarette smoke—came together to make our planet. In cosmically large quantities, dust blackens the Milky Way, blocking our view of most of the stars. And each star that dies rains more dust out into the galaxy, like a black firecracker. It is this dust of expired stars that will form the next generation of Suns, Earths, and other heavenly bodies.

And here on Earth we wouldn't want to do without dust. For starters, a clean world would be an oppressively *muggy* world. In the planet's water cycle, water evaporates off the oceans and lakes, condenses in the air, and falls back to the ground. But that condensation step assumes a sky full of dust, upon whose little surfaces water vapor can gather. Without dust, water vapor wouldn't begin to condense until the relative humidity was about *300 percent*. This would make the sultriest summer day seem dry and crisp by comparison. For lack of a more suitable nucleus, the water vapor would condense on your body.

Since a cloud is just a collection of water droplets condensed around various dusts, a shortage of dust also implies a shortage of *clouds* in the sky. And clouds reflect much of the sunlight that hits them, casting shade on the planet. At any given time, they cover about half the Earth. Without them, it would get mighty warm down here.

Many of the dusts that roam the Earth are little bits of life, whose ability to travel on the wind keeps the planet healthy and green. Fungi, for instance, make a living breaking down a variety of substances, including the dead flesh of plants and animals, and even rocks. Their efforts free trapped nutrients and enrich the soil. And the overwhelming majority of fungus species have adapted to fling their spores into the wind. These tough spores travel the world, falling back to Earth at the whim of wind and rain.

Many pollens also evolved to exploit the wind. The bigger grains hitchhike aboard bees and other nectar hunters. But the smaller ones sail through the air on their own, perchance to touch down on a suitable flower, thus ensuring the perpetuation of green and growing things.

Microscopic diatoms, which are glass-shelled algae, may distribute themselves this way, too. Even minuscule worms called nematodes are small enough to climb onto the wind and spread their race. Antarctica, for instance, was probably scrubbed clean of life in the last ice age. But now a variety of microorganisms, including the relatively large nematodes, have colonized the cold patches of dirt in the continent's McMurdo Dry Valleys. The most likely explanation for their presence is that their ancestors flew in from South America, Africa, or Australia.

Among the many marvelous subtopics of dust research, one that refuses to dry up and blow away is the notion that some tiny life forms not only ride the wind, but also reproduce in that dusty domain. Various researchers have proposed that some bacteria help water vapor to condense in the sky and then divide and multiply inside the drops they create.

Even the billions of tons of lifeless rock dust that clot the air downwind of deserts are valuable to the Earth. Certain islands in the Caribbean would be naked, gray rock if it weren't for the dusts of deserts and volcanoes that settle heavily upon them. Instead they are humps of lush and happy vegetation. Likewise the tapestry of the Amazon rain forest is indebted to dust. In such a rainy climate, water quickly flushes nutrients out of the soil. But each winter, when the trade winds head southwest from the Sahara, rich dust rains down on the South American forests and refreshes the soil.

Falling rock dust feeds tiny mouths in some of the world's most desolate places. On the Earth's glaciers, settling dust arrives like a catering service, spreading assorted dishes out for the enjoyment of some of the hardiest life forms we know of. Even *inside* a glacier we'll see that well-traveled dust can sustain a tiny web of life. Dust that falls in the ocean can also fuel a bloom of plants. These plants are microscopic phytoplankton. But despite their unobtrusive size, plankton are the bread and butter of the oceanic food chain. And in a twist on the "dust to dust" cycle, they sometimes take nutrients from falling desert dust and then send aloft a sulfur-rich dust that plays a key role in forming clouds.

To some degree, scientists have learned how an assortment of living and dead dusts tinker with the weather. And it's now becoming clear that dust alters the world's long-term *climate,* too. Traditionally, climatologists have focused their fears on *gases* that trap heat near the Earth, but as the globe gets warmer, little airborne specks have become a very big topic. Scientists now know that some of our dusts reflect sunlight and cool the planet. And others, especially our black soots, may be soaking up huge amounts of heat as they roam the sky. Some marvelous theories even implicate a global *blizzard* of dust in the sudden retreat of the glaciers at the end of the last ice age. But for now, the brightest minds on Earth can't say exactly what dust is doing to the thermostat, whether for better or worse.

The relationship between dust and humanity has also been complex—for thousands of years.

Eight thousand years ago, Chinese farmers discovered the charms of massive deposits of desert dust that had settled out of the air in central China. This blanket, about three hundred feet thick, was effortlessly tilled and nutritious for plants. Today similar dust deposits all over the world, including the central United States, are under intense cultivation. Unfortunately, as we'll see, loosening this ancient dust can sometimes be disastrous.

Perhaps four thousand years after Chinese farmers dug into their dust, people in ancient Mesopotamia were melting down their own local powder to manufacture rocks. At a site called Mashkan-shapir, archaeologists recently discovered large, flat rectangles of black rock, whose composition resembled no natural basalt. But the chemistry of the rocks did match the dust deposited on the banks of the local river. The archaeologists speculated that a natural shortage of wood and stone inspired the people of Mashkan-shapir to heat their dust to twenty-two hundred degrees, then mold the molten dust into rocks.

At that same early date, the people of Finland were exploring the merits of a special dust of their own. This dust, which was pounded from strange, fibrous rocks, strengthened the clay that they used for both pottery and house chinking. Farther south in Europe, people eventually learned to weave the fibers of this same rock—asbestos—into fireproof cloth. And early naturalists did notice that asbestos weavers were a particularly unhealthy lot.

On the other side of the world the Maya people at Tikal, Guatemala, seem to have carefully added a substantial percentage of volcanic dust, or ash, to their pottery to toughen it. That tradition demands a large supply of ash—which has produced a mystery: The nearest ash deposits aren't terribly close by. Was volcanic dust so valuable that it was worth lugging through a hundred miles of jungle? An alternative explanation is equally intriguing: Central American volcanoes were a lot more active a lot more recently than we realize, throwing their ash all the way to Tikal.

Today humanity still employs dust for planting crops, for building, and for pottery—and for thousands of other purposes. Cement walls are a mixture of rock dust and pebbles. Sheetrock is a mineral powder, compressed into a convenient form. Colored dust gives paint its hue. Rock dusts give scouring powder its grit, toothpaste its polish, and talcum powder its silkiness. Eye shadow can be a mixture of dazzling dusts, from talc to powdered fish scales and pigments. Aspirins and vitamins are compacted dusts. Magazine paper is made shiny with the thinnest coating of dried clay dust. Pencils hold a core of pressed graphite powder. Bread is made of powdered wheat kernels, and so is pasta. Yellow mustard is the dust of mustard seeds, and soft cocoa is the dust of hard cocoa beans. Modern life relies heavily on dust.

One reason we powder so many things is that dust offers huge amounts of surface area to work with. Since chemical reactions generally take place on the surface of an object, the more surface you can provide, the more intense the reaction will be. Imagine, if you will, steeping fifty whole coffee beans in a mug of hot water. Yuck. Then imagine grinding fifty coffee beans to dust and repeating the experiment. Or imagine dropping a bar of solid soap into the washing machine with a load of clothes. Then imagine shredding that soap into powder and repeating the exercise. More surface area permits more interaction.

This characteristic can produce results both wonderous—and woeful.

Some of the dusts that swirl around us are fearsome and invisible rogues. Leave aside for a moment poison particles launched by human industry. Plain old desert dust has a dark side of its own.

Seventy-five million years ago, for instance, simple desert dust seems to have set a subtle trap for a fieldful of dinosaurs. One minute these formidable creatures were going about their domestic duties. And the next minute the dust in the surrounding sand dunes conspired to entomb them. (The detective work required to reconstruct such an old murder scene, and to implicate something as easily overlooked as dust, is considerable, we'll see.)

Perhaps those dinosaurs were the lucky ones. Ten million years later the dinosaur story would come to a much slower and more final close, as a worldwide cloud of dust from a giant meteorite impact darkened the sky and blotted out the Sun. That dust murdered birds, sea life, and small, pioneering mammal species as well.

Desert dust still delivers trouble to this day. A dust-related disease stands charged with the slaying of purple sea-fan corals. Dust from the Sahara has long beaten a path across the Atlantic Ocean to rain down on the Caribbean. But in the 1970s a terrible drought in the African Sahel region south of the great desert began to send extra dust rolling down this skyway. And as the falling dust grew thick in the Caribbean in the early 1980s, scientists saw a plague sweep through the coral reefs. Coincident with the dust invasion, two species of coral were nearly wiped out, a species of sea urchin was decimated, and the purple sea fans developed dark, lumpy lesions. It took some sleuthing, but a scientist has pinned the sea-fan plague on a fungus in the Saharan dust.

Scientists are now examining this dust more closely and finding everything from radioactive elements to mercury and an impressive array of fungi. In the summer in southern Florida, says one longtime dust scholar, this far-flung desert dust is the most common sort of particle in the air. There may be implications for human health.

Health experts already know that some dusts can be deadly to people. When they rank U.S. cities by the quantity of pollution dust in the air, then rank the same cities by deadliness, they find a match. The dustier the city, the higher the death rate. One federal agency estimates that pollution dusts alone kill sixty thousand people each year in the United States. The crucial question in this case of mass murder is . . . *which* dusts do the killing?

Some dusts are *obviously* fatal. Coal dust, for instance, kills 1,500 miners every year in the United States. The dust of powdered quartz kills 250 more miners, sandblasters, and other laborers in this country. The needle-shaped dusts of asbestos cause deadly cancers of the lung and gut. But none of those dusts hangs terribly thickly in city air. Something else is at work. The clues are piling up against the tiny chemical dusts of our own making.

The dusts we find indoors can be as kind and as cruel as those found outside.

The dust bunnies that skulk beneath the couch and behind the refrigerator contain everything from space diamonds to Saharan dust to the bones of dinosaurs and bits of modern tire rubber. But they also hold poisonous lead and long-banned pesticides, dangerous molds and bacteria, cancer-causing smoke particles, and a sample of all the convenient chemicals that we innocently distribute through our houses in the name of cleanliness. The dust bunny is riddled with allergy-inducing dust-mite parts, with the mites themselves, and with the predatory mites and pseudoscorpions that stalk and kill them.

In addition, house dust bears some blame for lead poisoning among children. As children crawl across carpeting—especially aged, dust-packed carpeting—their sticky little paws gather dust. And then those paws go into their mouths. One of the best indicators of how much lead a child's blood will contain is how much lead a sample of carpet dust contains.

Oddly, if it weren't for the chemicals and metals that foul house dust, we might learn to *love* our dust bunnies. For decades, allergists have shot up some of their patients with a distillation of dust taken straight from the vacuum-cleaner bag. Although the secret to the success of this bizarre protocol isn't

known, allergists swear it does tame dust allergies. And some of the most rivet-
ing research in all of dust science is now drawing a connection between dusty
homes and *healthy* children. An epidemic of asthma is exploding among the
children of developed nations. But a flurry of studies is showing that babies
who do their crawling and finger sucking in dusty, germy houses are less likely
to get the wheezing disease. Something in house dust, doctors insist, must
toughen a baby's immune system.

Indoors or outdoors, dust is unavoidable. And a marvelous fraction of that
dust holds the secret to our past.

Some of the dust that swirls around us was knocked off distant, colliding
asteroids eons ago. Some of it boiled off comets that may have passed our way
a few years ago or a few centuries ago. This stuff, still holding its ancient grains
of primordial stardust, settles on Earth at a rate of one magical speck per
square meter per day.

Because these extraordinary dusts carry the secret of our cosmic past, we'll
see that scientists go to extreme measures to capture them. And catching these
microscopic time capsules is only half the battle: To analyze such smoke-small
specks is sometimes simply impossible. But whenever a dust scholar is able
tease out the chemical fingerprint of a grain of space dust, she comes a little bit
closer to understanding the origins of our world.

That is the secret of our past.

The secret of our futures—our personal, *individual* futures—is also circulating invisibly beneath our noses. Just as the dust of dinosaurs now roams the
air, so, too, will the dust of a decayed *you*. If your body is buried, it will eventually merge with the surrounding soil. Then hundreds, or even millions, of
years later, erosion will open your grave and scatter you around the world. If
you are cremated and scattered, your path to dust will be much accelerated.

Even the most heroic efforts that some people now undertake to evade the
dust state won't forestall the inevitable. Even if a body lasts until the end of the
Earth, dust is in its future. The hands-down odds are that the Sun's slow demise in a few billion years will have the side effect of roasting our planet. The
puff of smoke that was our world will blow on the solar wind, out across the
dusty galaxy.

2

LIFE AND DEATH AMONG THE STARS

In the year 1054 a Chinese astronomer, perhaps standing atop a tall stone observatory, looked skyward and gasped with surprise. It was midsummer, and it was daytime. But there, in the blue ceiling of the world, blazed a reddish-white star. And when night fell, the peculiar star raged with even more fury. Chinese astronomers, equipped with intricate sky maps and interlocking metal hoops that mimicked the rotating heavens, had been faithfully noting the appearance of such "guest stars" for more than a thousand years. They sensed that these bright vagrants that wandered across the starscape, or bloomed suddenly and faded away, bore some crucial message. But what was it? Following tradition, the Chinese observer interpreted the star as best he could.

"I humbly observed that a guest star has appeared," began his note, which concluded with an interpretation that would flatter the Emperor. ". . . the fact that the guest star does not trespass against Pi and its brightness is full means that there is a person of great worth."

But even as the ink was drying on his report, the guest star was dimming. By autumn its fire no longer burned through the daytime sky. And by the following autumn the star was hard to see in the night sky, too.

And so the astronomer turned his attention to brighter objects, not realizing that one of a guest star's most profound messages lies in its very dimming: the guest star was shrouding itself in black dust.

This cast-off dust, we now know, gathers into enormous wombs that gestate the next generation of stars. Remnants of those dust wombs roll themselves into rocky planets like the one beneath our feet. And, most eerily, these specks, which twinkle with everything from sugar to diamonds, may be the tiny factories in which nature builds molecules with the ability to spring to life.

In this dust we are finding our origins.

. . .

Nearly a millennium later a Dutch astronomer blew the Earthly dust off the Chinese astronomer's records. He plotted the location of that momentous guest star. And when he pointed his telescope, he found himself peering at the well-known Crab Nebula, which lies in our own Milky Way galaxy, about 7,000 light-years from Earth. He concluded that the guest star had been a supernova, the explosion of an uncommonly large star.

This star had a glorious, if brief, life. It began as a massive ball of simple gases, spiked with a pinch of dust—only nine elements in all. But after just a few million years, the star's central furnace burned through most of its fuel. The star shuddered and shook off its outer shells of gas. Shortly thereafter the furnace itself exploded, launching a shock wave that rippled violently outward, overtaking the fleeing gas shells. And the fury of that shock changed some of the gases into strange new atoms—platinum, gold, titanium, and uranium among them.

In the months after the explosion the concentric shells of mixed vapors—glowing green, blue, and red—broke into irregular pieces. The pieces, still hurtling through space at millions of miles an hour, shrank into discrete clouds. And as the vaporized elements cooled, they condensed into tiny grains of dust. Silicate dust, chemically akin to glass. Gold dust. Radioactive uranium dust. Dusts of europium and platinum. And dust of diamond.

Admittedly, these last items were a few carats shy of museum pieces. Since carbon isn't plentiful in the shells of an exploded star, the microdiamonds would have had to grow rather quickly. They may have gathered a new carbon atom every five hours for a few hundred days following the big explosion. Then carbon atoms would have grown scarce, frustrating the ambitions of the little jewels. One astronomer has proposed that if bacteria required engagement rings, these space diamonds would do nicely.

But even as the diamond dust peaked, other dusts continued to condense and accumulate. Each day the thickening dust blotted out more of the guest star's shining wreckage. A few years after the explosion virgin dusts resembling powdered graphite and glass swirled like dark fog. This wasn't teeth-gritting stuff: two hundred individual grains, shoulder to shoulder, would amount to the width of one human hair. And the diamonds were far too tiny to shine. Nonetheless, this dust faced a dazzling future.

Today the Crab Nebula resembles a translucent green egg, shot through with red filaments. And it is darkened with clots and ropes of dust. Although its dust-making days are probably finished, the Crab's dust-*distributing* era has just begun, says Kris Davidson, a University of Minnesota astronomer who did pioneering research on the nebula's composition.

"The rule of thumb is as soon as you get below a thousand Kelvins [1,300 Fahrenheit], dust automatically forms," Davidson says. "My guess is the Crab was a very hot nebula. So dust probably didn't begin to form until a few years after the explosion. And after a hundred years there would have been *significant* dust."

In the thousand years since that explosion the Crab Nebula has swelled like a dust-spotted balloon, reaching 70 trillion miles in diameter. And it's not nearly out of steam, says Davidson.

"Normally, supernova remnants stop when they run into other clouds of gas," he says. "But the Crab is all alone, five hundred light-years above the plane of the galaxy. So one half of it is going to keep expanding into intergalactic space. The other half might collide with the plane of the galaxy in thirty thousand to a hundred thousand years."

On human time scales glamorous guest stars like the Crab are rare. Only one in a thousand stars in our galaxy is big enough to expire in supernova style. And although supernovae are glamorous dustmakers, they're not very practical dustmakers. Their dusts are poor in carbon, nitrogen, and some other basics on which life depends.

But from the slow perspective of the cosmic clock, both these giant stars and their humbler siblings are constantly blowing dust bubbles in the Milky Way. Between a hundred billion and a trillion stars of all sizes inhabit a typical galaxy, and nearly all of them fling a bit of dust out into space when they die. Sometimes even a healthy middle-aged star will emit an isolated cough of dust from its hot atmosphere before settling back into its clean-burning routine.

A modest star, like our own Sun, is so frugal with its fuel that it might burn for 10 billion years. But then it, too, will swell and throw off hot shells of fresh-minted atoms. With little fanfare, the humble and abundant stars like our own cook up much of the universe's carbon dust—as well as elements like strontium, yttrium, barium, and aluminum. In fact, the bacterium that required a birthstone ring, instead of a diamond, would do well to rummage through the death dust of a more modest star. There it would find abundant specks of aluminum oxide, which, depending on its impurities, is known on Earth as ruby or sapphire.

Even the charred *core* of a deceased star can produce dusts we couldn't live without. From time to time one of these weighty "white dwarf stars" will get close enough to a companion star to begin stealing away its gas. When the thief eventually becomes overburdened, it will explode. And when the glowing gas remnants cool, the darkening dusts will be rich in such meat-and-potatoes elements as iron and its relatives—chromium, manganese, cobalt, and nickel.

Today the Crab Nebula has likely made all the flashy dust it can, Davidson says. Now the newborn dust is hurtling outward, proceeding to its next incarnation.

Scientists can predict what the Crab Nebula's dust is made of and how fast it's rocketing away from its birthplace. But from Earth, many trillions of miles away, there are still some details about this dust that they can't quite detect. One thing they don't yet know is what a piece of space dust *looks* like.

"Dust is not well defined," says Steve Beckwith, an astronomer who studied space dust before becoming director of the Space Telescope Science Institute, which decides where to point the Hubble. He's a tall man, with unusual gold eyes. "We don't know the precise shape or composition of the grains," he says. "Is it little spheres? Is it strings? Plantlike things? A lot of it is probably little snowflakelike things."

Beckwith's is a partial list. One school of thought says that the average piece of space dust is a naked grain of pale, glassy stuff—or of black graphite. Another school proposes a glassy core, coated over with intriguing, organic molecules. Others predict a fluffy mixture of glassy bits, carbon, ices, and organic molecules. The most freethinking school holds that space dust is so thoroughly organic that it's downright biological, bursting with potential life.

Astronomers also have trouble pinning down the *size* of the Crab's hurtling dust. Judging from the radiation that space dust sheds, they can say how small it is: It's smoke small. But how big does space dust grow? The radiation is mute on that subject. As one dust scholar has complained, "Some of the dust really could be the size of Toyotas, and we wouldn't know it."

But a portrait of the average dust grain is slowly coming into focus. The same space shuttle that carried John Glenn aloft in 1998 also carried a space-mimicking canister. The canister was injected with a puff of carefully crafted dust grains. And as electronic monitors looked on, the microscopic beads began linking themselves into short, branching twigs. Unfortunately, the loose dust rather quickly migrated to the sides of the canister, and stuck there, before the twigs could grow very large. In a future experiment, the canister will tumble like a clothes dryer, to discourage the dust from clinging.

· · ·

Whatever this fresh baked dust looks like, more of its clutters the galaxy every day, as each dying star converts a little more gas into heavy elements. So space is not empty. It's very, very dusty.

If you journeyed into deep space, marked out a hundred-yard-wide cube of space, and counted the dust grains therein, you might find just twenty microscopic grains. But to get a sense of scale, the skimpy gap between Earth and Pluto alone amounts to about 63 billion such cubes. At twenty dust grains per cube, the specks add up. Wherever astronomers point a telescope, they are met with a mist of dust.

David Leckrone is NASA's project scientist for the Hubble Space Telescope. When the school-bus-size telescope first opened its eye on the universe, it was confronted with the same plague of dust that blinds Earth-based instruments. Leckrone, perched on a corner of the desk in his Maryland office, is a quick, catlike man, with silver hair. He proffers a photograph of our home galaxy, the Milky Way. It is a view toward the center of the galaxy, taken from the location of our own planet, near the outer rim. And it's dark and moody. The stars blaze only in pinpricks and fat clusters of light.

"This would be a solid sheet of light," Leckrone says, admiring the photo with a frustrated grin, "if you could take out all the dust."

All that dust is the accumulating wreckage of stars like the Chinese guest star. And the story of its future is also the story of our own world's past. So let's spin the cosmic clock *backward* a few billion years and focus the telescope on the cloud of dust that built our solar system.

Six or 8 billion years ago, long before the Earth was born, there were guest stars aplenty. No eye witnessed their brilliant visits. No inky brush laid out a description of their location in the Milky Way. No one was around to venture a guess as to their meaning.

As cosmic time passed, the guest stars flared and faded like a slow-motion display of fireworks. Eon after eon they, and the dimmer, less ostentatious stars, steadily churned out dust.

And one fine cosmic afternoon, roughly 5 billion years ago, the violent galactic winds swept a little too much of that dust into one pile. Although the fireworks continued to send storms of destructive radiation through the galaxy, this dust pile was so immense that it withstood the battering. In its darkest interior the ongoing fireworks could scarcely be heard.

It was one of many such dust clots in our galaxy, which to this day is studded with them. The constellations Taurus and Orion each host a massive black cloud. All told, the Milky Way is home to about four thousand giant dust clouds, and many more smaller clots. We now know that these dust piles are the largest objects in the galaxy. Some of them span 300 light-years—or about seventy times the distance between us and the nearest star. Some hold enough gas and dust to produce a million Suns. And through their study of these clouds, today's scientists can decipher the story of our own birth cloud.

The size of our ancestral dust cloud remains a mystery. But when it was first pushed into a pile, it would have been rather anemic, poor in both gas and dust. It held less than one-trillionth as many gas atoms per cubic inch as does the air we breathe on Earth. And even these sparse gas atoms grossly outnumbered the dust specks.

That something as gigantic as the Sun could grow from this thin soup seems ludicrous. But the cloud was *inconceivably* huge. And although each dust grain was the size of a grain of smoke, taken in huge quantities they were a force to be reckoned with. The storm of radiation that harassed the galaxy ran up against this dust and stopped.

Ultraviolet radiation was especially perilous to our ancient dust cloud, says astrophysicist David Leisawitz. Leisawitz, whose office lies across the Maryland NASA campus from Dave Leckrone's, is a textbook example of his breed: he travels a tight orbit around a coffeepot, his sentences stagger under a burden of terminology, and his walls are papered with pinups of satellites.

All the stars around our ancestral dust cloud were shedding ultraviolet radiation, or UV, into space as they burned, Leisawitz explains. This radiation had a special ability to break apart the dust grains and complex molecules that were essential to building our world. But in our cloud the dust grains repelled the UV assault.

"A dust grain is like a little rock," Leisawitz says, sketching a dust speck and then directing a menacing and wavy line toward it. "It absorbs almost all the UV that strikes it." In our old dust cloud the outmost dust grains took on the role of armor.

Heat is another foe of would-be worlds, says Leisawitz. When our parental dust cloud first formed, the gas atoms inside it were still blazing hot from their supernova birth. The heat of nearby stars may have piled on a few more degrees. Unless this gas cooled off, it could never gather into balls.

Leisawitz draws another wavy line, this one traveling *away* from the dust speck.

"Dust absorbs UV," Leisawitz says. "But it *emits* IR." IR is infrared radiation—the heat that you see rippling off hot pavement. In plain old English the dust in our ancestral cloud acted like a zillion little radiators, throwing heat clear out of the womb.

Slowly, the interior of this dust cloud became one of the coldest places in the universe. A trickle of radiation left over from the Big Bang heated every-thing to a minimum of 3 degrees above absolute zero. And a handful of UV and cosmic rays slipped past the outer guard of dust, adding a few more degrees. So our dust cloud's center hovered at a nippy 441 below zero Fahrenheit.

The soothing womb of dust calmed and cooled its innards until things began to stick together.

Inside the cloud the dust grains now acted as laboratory benches, where atoms could meet each other and grow into marvelous molecules. But even with the aid of dust the molecules that surround us today took ages to grow. For a sin-gle atom of gas to encounter one grain of dust may have taken a million years.

Nonetheless, atom after atom landed on a speck of cold dust and awaited a mate. Some of the molecules that resulted left, to build bigger molecules in the dark and ripening dust cloud. And on the surface of the abandoned dust grain, fresh atoms built new molecules. Some of these nestled tightly against their host.

The dust evolved. Water molecules froze on its surface. Nitrogen and other gases froze there, too. Occasionally, an infiltrating beam of UV slammed into those ices, kicking off a chemical reaction that built even larger molecules. The birth cloud of our Sun and planets grew more sophisticated.

Just *how* sophisticated, is a subject of debate. The simplest molecules, like a rudimentary sugar that astronomers discovered in a dust cloud in the summer of 2000, are the easiest to identify and confirm in the laboratory. Bigger mole-cules are tougher.

"A molecule containing eleven atoms is the biggest that's been *confirmed*," says Emma Bakes, an English-born NASA scientist who tries to identify the "fingerprints" of individual chemicals, based on telltale radiation that escapes from dust clouds. "We feel quite sure there are molecules with *eighty* carbon atoms. But confirmation is terribly difficult." Sorting out the individual fin-gerprints can take years.

Max Bernstein, a colleague of Bakes's, takes a different approach: he re-creates an experimental space-dust cloud in his laboratory and studies the strangely familiar molecules that appear. He applies a coating of ices to the surface of man-made dust, then subjects this coating to the same UV radiation that would have occasionally penetrated the womb of our ancestral dust cloud.

"The radiation breaks the bonds of the molecules," he reports, "but because they're frozen in ice, they can't go anywhere. Then, when they warm up, they reconnect with other fragments, making much more complex molecules." Ketones, nitriles, ethers, and alcohols have emerged from the high-tech dust cloud, along with a carbon-rich molecule that produces amino acids when it's dunked in warm, acidic water. Traditionally, scientists thought that amino acids could be made only in liquid water. But Bernstein's experiments are indicating that in the absence of water, icy dust will do. "We're getting extremely encouraging results," he says cautiously. "We *are* seeing amino acids."

A third NASA colleague, Jason Dworkin, is working with space-dust molecules that quickly assemble themselves into hollow, watertight spheres when he drops them into water. "All life lives inside of membranes," Bernstein says. "A membrane a very desirable thing."

The notion that a special group of dust-forged molecules could behave like cells—could somehow *become* cells—has implications for our world and for others. If these talented chemicals formed in our ancestral cloud, they could form in *every* space-dust cloud. So wherever there is dust in the universe—and dust darkens every galaxy—there is also the potential for life.

As the dust grains grew into loose collections of mineral and metal bits, strange molecules, and various ices, the explosion of any nearby star could have blown a hole in the womb. Even the violent winds shed by a nearby star could have eroded it. Had the growing dust grains been exposed, they would have been stripped of their chemical coatings and set adrift. It's a common occurrence. It has been calculated that any given speck of space dust may join ten different dust clouds, over a billion years, before it is finally swept to the center of a cold cloud and caught up in the whirl of star birth. But this time the dust hung together. And every atom of everyone on Earth was preserved.

Even as today's chemists read out our cosmic history from the surface of space dust, astronomers are chiming in, too. With telescopes they have caught dust in the act of giving birth. And from watching brand-new stars struggle out of dusty wombs, they can now tell the tale of our own Sun's birth.

"Dust is the environment of stars," says Hubble scientist Dave Leckrone. "Stars are born in the dark."

Setting aside the portrait of the dusty Milky Way, Leckrone proffers an old Hubble photograph of a corner of the Eagle Nebula, a vast cloud of dust 7,000 light-years from Earth. Backlit by faint starlight, the nebula looks like the fingers of an inflated rubber glove, smoky-black with dust. On close inspection a few dull red cinders glow in the murky fingers. They are young stars, still enveloped in heavy dust wombs.

Without a thick cloud of dust a star cannot form. Unfortunately, those same wombs render normal telescopes powerless to spy on embryonic stars. It wasn't until the 1980s that scientists solved this problem. Unlike visible-light waves, the longer waves of *infrared* radiation weave right through the dust. When astronomers adapted telescopes to collect infrared, they built themselves stellar sonogram machines.

Leckrone triumphantly holds up two photographs of the Orion Nebula. In the first image, one of Hubble's early cameras shows the nebula as opaque as swirling fog. But in the second, taken by an infrared instrument installed in Hubble by astronauts, the nebula is a sparkling nest of brand-new stars.

Just two centuries ago dust clouds were dismissed as "holes in the heavens." Now Leckrone and his colleagues fondly refer to black dust clouds as "stellar nurseries." And they can assure us that our own Sun and Earth took shape in the same sort of dusty envelope.

In the cold center of our vast parent cloud the first hint of the Sun would have been a faint thickening of the gas-and-dust soup. For a sense of the size of this solar seed, imagine that our modern Sun is a grain of sand. By comparison, the thickening womb that gave birth to it was nearly a mile wide. The seed of the Sun slowly contracted, condensing the soup it held. And the more the seed shrank, the harder gravity tugged at it, shrinking it further. When the gases warmed under the pressure and threatened to expand, the dust radiated away the heat. After a scant million years the core was so dense that gravity suddenly got the upper hand. The core collapsed into a ball. The outermost layers, however, held their ground, forming a thick, dark shell around the new "protostar." Had an astronomer been watching, 4.5 billion years ago, this monstrous black shell, about a light-year in diameter, is all she would have seen.

But deep inside its dusty case the cool and sluggish infant Sun rotated just once every few million years. As gravity continued to squeeze the ball smaller, it spun faster, like a figure skater pulling in her arms. When the relentless pressure heated the gas faster than the dust could cool it, the Sun began heating up to shining temperature.

And the dust-womb remnants? Astronomers have no idea if our ancestral dust cloud was huge or modest. It may have spawned dozens of siblings for the Sun, which have wandered away in the intervening years. It may have cooked up one massive sibling that promptly went supernova, blowing our shared parent cloud to bits. We may never know.

As for the small shell of dust that once hid the Sun, some of that dust blew out into space when the Sun finally caught fire. But a sprinkling of it remained, circling the new star in a wide, doughnut-shaped ring.

. . .

The grains of dust in this doughnut had grown larger as the Sun took shape. The dark powder of microscopic rocks, metal flecks, molecules, sapphires, glass, and diamonds all roiled around the Sun, suspended in an orbiting river of gas. At the inner edge of the doughnut, nearest the Sun, each cubic inch of space was packed with billions of dust grains and gas molecules.

By the cosmic clock this bounty would be transformed into planets in a wink. But from the perspective of one dust grain it was a long and violent struggle. Early in the planet-making process the dust whirled wildly, like pebbles in a tornado. Many of the grains collided so violently that they bounced apart. Eventually, two grains going in the same direction managed to brush each other gently, and they bonded.

Doubled in size, this new dust grain tumbled on through the doughnut. All around, particles were crashing, rebounding, crashing, bonding, and growing. Onto the double dust grain froze a speck of water. Another grain of dust joined the pair. And then another. When the growing dust ball had gathered 100,000 individual grains, it was on the verge of visibility: it measured a whopping one-tenth of a hair's width.

This promising grain was growing steadily when the young Sun threw an adolescent tantrum, and a blaze of heat flashed through the doughnut. Some of the dust melted right down to vapors. Other specks half melted, then hardened into "chondrules," glossy red-and-black beads about the size of sugar crystals. The crisis quickly passed, and chondrules and dust alike resumed their mergers.

Within a hundred years our ambitious dust grain had grown a yard wide. As this boulder and countless others like it continued to sweep up dust, more sunlight filtered through the haze. The dusty doughnut flattened into a disk. Nearest the Sun the dust boulders flocked. Toward the middle of the disk, gases clumped and coagulated into the giant gas planets. And at the cold, outer edge of the disk, dust mixed with copious ice crystals to form a horde of small dirt balls—the comets.

Back in the rocky-dust zone, our dust boulder merged messily with other boulders. When it reached a mile in diameter, the Earthlet was officially a "planetessimal." Now it encountered other planetessimals only about every thousand years, but their similar orbits and their gravitational attraction improved their odds of merging, instead of shattering. That's not to say these mergers were gentle. They were powerful enough to melt and mix great patches of fluffy dust grains. The minerals and metals rehardened with the familiar texture of solid rock.

After 20,000 years hundreds of Moon-size bodies peppered the inner solar system, and the Earthlet was among them. After 10 million years only a handful of colossal balls remained circling the Sun. The Earth, Mars, Mercury, and Venus greedily cleaned up any remaining planetessimals, leaving only a band of stony asteroids orbiting between Mars and gaseous Jupiter.

The space dust that formed the Earth was no longer recognizable. Dust grains born a billion or more years before had now been melted down by endless crashes and by radioactive atoms that were still "hot" from their traumatic supernova birth. As the grains melted, some of their constituents separated and segregated themselves: much of the iron and nickel dust sank to the Earth's center. The lighter, rocky dusts rose and cooled into a crust. The *truly* light elements—the ices, the carbon, and even the marvelous lifelike compounds—collected atop the crust. Some of these precious molecules and gases may have blown clean away on the new solar wind.

One school of thought says that so much of the Earth's lightest ingredients blew away that life couldn't have arisen without a last-minute delivery of unmelted dust. Fortunately, way out on the rim of the old dust disk the comets circled. Much farther still, a huge population of Oort cloud comets may also have waited for a passing star to nudge them Sunward. And all these comets appear to have held their rich, organic dusts safely in cold storage. Analysis of the occasional ice ball that flies past the Sun suggests that comets contain a huge number of organic molecules, which may comprise nearly one-third of their bulk.

Even the rocky asteroids, dust clumps that formed much closer to the roaring Sun, may have managed to preserve some of the water and organic molecules of their original dust grains. Water, and signs of water, have recently been discovered in the stony innards of fallen meteorites.

And so, this theory goes, after the molten Earth had cooled, a blizzard of incoming comets delivered their fragile contents to the planet. Water, gases, and life-inspiring molecules pelted the newly hardened ground. Earthly molecules and heavenly ones then conspired to produce life.

Even free-floating dust grains may have been an important source of the precious molecules, proposes NASA dust chemist Max Bernstein. "Some interplanetary dust particles are fifty percent organic material by weight," he says. "No one has ever pulled amino acids from dust, because the grains are so freakin' small. But we know they're in meteorites, so let's assume they're in dust particles, as well. And since *tons* of these dust particles fall to Earth every day, they add up."

A laboratory simulation indicates that between 1 and 10 percent of the crucial, lifelike molecules could survive the rigors of atmospheric entry and impact.

Like a haunting reminder of the debt we owe to dust, a wispy disk of it still circles the Sun. On rare occasions you can actually see a glowing slice of this "zodiacal light," which the Italian astronomer Giovanni Cassini described in 1683 as a glowing triangle that hovers above the horizon on certain dawns and evenings.

Most of this zodiacal dust is not so much newly minted as freshly grated. Asteroids are probably dust balls that lost the race to reach planet size. But if they lagged in gathering dust, they now excel in sprinkling it back into the solar system. The gravity of their enormous neighbor Jupiter disturbs their orbits when it rolls past them. The jostled asteroids then collide with each other. And each crash releases dust into the solar system.

Various moons may also contribute to the disk when they're hammered by meteorites. Mars's moon Phobos is three feet deep in dust, thanks to the rain of meteorites it suffers. The faint rings around Jupiter have been identified as more moon dandruff. One of Jupiter's moons, Io, is suspected of belching volcanic particles, which the big planet's magnetic fields whip into the solar system at astounding speeds. And even our own Moon was once suspected of harboring great drifts of dust, a possibility that gave planners of NASA's first moon-landing mission chills: suppose the triumphant astronauts touched down, only to sink out of sight?

Comets also sprinkle dust when they visit the solar system. Comet Linear, as it sped toward the Sun in the summer of 2000, began to thaw as it passed Jupiter. Sunlight vaporized the snowball's outermost ices of water, methane, and ammonia. This set free ancient grains of dust. Then, as Linear cut a close turn around the Sun, violent explosions popped apart the comet's very nucleus. Huge bursts of bright dust appeared as Linear was split into at least a half dozen mini-comets.

Even the occasional grain of *foreign* dust wanders through our neighborhood. The galaxy is packed with dying stars, and their dust flies in all directions.

Yes, if space is dusty, the solar system is positively choking on the stuff. Because each little grain spirals slowly into the Sun, the dust never gets thick enough to blot out the light. But there is always enough of it to see, if you know where to look.

The hazy triangle of the zodiacal dust is easiest to spot from the dark countryside, around the spring or autumn equinox. Then the sunlight that sweeps past the Earth lights up a wedge of particles that points toward our sibling planets. In spring, the arrow glows on the western horizon, after sunset. In autumn, it's on the eastern horizon, before dawn.

Curiously, the zodiacal light is too large and too subtle to view with a telescope and is best studied with the naked eye. So, had the Chinese astronomers of a thousand years ago wished to ponder the cosmic implications of space dust instead of glamorous guest stars, they would have been ideally equipped.

3

A LIGHT AND INTRIGUING RAIN OF SPACE DUST

The Earth is still gathering more than a hundred tons of space dust every day—to the delight of scientists. Each speck, broken from an asteroid or shed by a comet, might hold a hundred thousand smaller specks. And among those sub-specks are the diamonds and sapphires, the inky-black carbon and rich organic molecules that date to the days of our planet's creation—and even earlier. The very oldest grains buried inside a piece of space dust can tell stories about the long-gone stars that produced them. The chemicals that sometimes en-wrap these grains whisper about the conditions inside our ancestral dust cloud. Even the manner in which the tiniest grains cling together holds clues about how the dust grew as the Sun and planets took shape. "Since every atom in our bodies came from inside of stars," explains astrophysicist Don Brown-lee, "by studying these interstellar dust particles, we can learn about our cosmic roots."

Brownlee is a space-dust pioneer who has recently gone so far as to send a spacecraft out to capture this whispering dust. With a floppy haircut and a zippy green canvas shirt, Brownlee is a little youthful to play the Father of Cosmic Dust. On his wall is a poster of an astronaut in a space suit, standing on the Moon and taking a pee. But Brownlee cracked open the subject with his Ph.D. research in the 1970s, and he now surveys the field from a modest office at the University of Washington in Seattle.

The dust that he studies is a hundred times bigger than the first generation of dust that darkened the early solar system. But it is still fiendishly difficult to work on. Although there is always a bit of space dust in the air around us, it's nearly impossible to isolate it from the clutter of Earthly dusts. Even when scientists do catch a dust speck, they're faced with the challenge of storing and handling an item that's one-tenth the width of a human hair. And then the task of determining the celestial source of the 100,000 individual micrograins that make up a single mote could keep an astronomer busy for a lifetime.

If the mysteries are overpowering, they are also entertaining. Brownlee is prone to hunching his shoulders and saying things like "We don't have any idea. Isn't that *great?*"

Happily, scientists receive a steady supply of clues. The Earth grows fatter every day, snowed under by a continuous microscopic flurry of space specks. Rare as they are, on average, every square yard of the planet should nonetheless receive one speck each day. Statistically, it's a good bet that there's a fresh piece of space dust on the hood of your car daily—and a dozen fresh specks on your roof. Lie on your lawn for a day and you stand a shot at being pelted by a glassy mini-marble or a delicate crumb of comet dust.

"They're everywhere," Brownlee says. "You eat them all the time. Any carpet would have 'em."

Our planet orbits more or less inside the disk of zodiacal dust. About three-quarters of the dust surrounding Earth is cast off by colliding asteroids as they orbit out beyond Mars. One-quarter of the dust gushes from comets as they sail Sunward and begin to thaw. And except for the tiniest specks, all this dust is obliged to spiral toward the Sun—which means much of it must pass the Earth.

For a small grain of dust to migrate from the asteroid belt to the Sun takes about 10,000 years. For a large grain, pursuing its orbit with more determination, the fateful spiral can last 100,000 years. Most of the spiraling specks, of course, are not intercepted by the Earth, and they eventually meet a violent end on the Sun's doorstep—although exactly what the end looks like isn't clear.

"Are they vaporized? Do they boil down and then blow back out?" Brownlee queries. "I don't know. It's wonderful! I *think* I know what happens," he continues with a bashful laugh. "We've seen some most peculiar structures." Like Mickey Mouse spheres.

"If a piece of dust doesn't collide and get broken up, it might get hot enough to melt," Brownlee explains. "The silicate forms a sphere. And then the metal forms little beads on the outside. They look like Mickey Mouse." And these blobby little dusts are so small that they no longer spiral in. Instead, they blow back out, taking another chance of being swept up by the Earth as they pass. Each second of the day a few more tons of space dust get so close to the Sun that they melt down and blow back out on the solar wind. But as quickly as the Sun cleans up, new dust arrives. Asteroids continue to bang heads, and comets continue to leave dusty trails across the solar system.

And that's in the "clean" times. Occasionally, the amount of dust spiraling past Earth reaches a mysterious crescendo. Core samples of sediment punched from the sea floor show that every 100,000 years, the thickness of the Earth's dust-fall triples. So in about 20,000 years, we may wake up to find the Sun dimmed by dust.

There are competing theories to explain the onslaught. One is that the Earth's orbit tilts slowly in and out of the zodiacal dust disk. When we orbit deep inside the disk, we get heavily dusted; when our orbit tilts out of the dust disk, the flurry relents.

A competing theory lays the blame on the shape, rather than the angle, of Earth's orbit. Over the course of 100,000 years our planet's path around the Sun shifts slowly from a more circular to a more oval orbit. In the more circular phase, the planet travels more slowly and could possibly sweep up two or three times more dust than it would during oval phases.

A recent computer-model experiment suggested that it is indeed the shape of the Earth's orbit that matters. But scientists in this "circular orbit" camp are still hounded by an Earthly contradiction in the data: sea-sediment cores stubbornly insist that cosmic dust was thinnest just when the computer says the Earth should have been gathering dust by the armful. And when the planet ought to have been speeding *past* the cosmic dust, cosmic content in the sea mud hits its peak.

Although mysteries remain, this work may nonetheless have put to rest a theory that blamed storms of space dust for the Earth's rhythmic ice ages: even during the dustiest times of the cycle, there's simply not enough dust to dangerously dim the Sun.

A related theory proposes that the ice ages were brought on by cataclysmic asteroid accidents. While the 100,000-year cycle merely triples the dust, one big asteroid crash might add three hundred times the normal amount of dust to the disk. That would dim the Sun noticeably. And it could conceivably fling chunks of broken asteroid out of their orbits and send them Earthward. It's a tidy scenario that could account for periodic outbreaks of extinction on Earth: you chill 'em with dust, and you pound 'em with asteroids.

But close scrutiny of fallen dust suggests comets, not asteroids, may deserve this devastating reputation—at least for one event that deeply scarred the Earth. Ken Farley, a geochemist at the California Institute of Technology in Pasadena, has discovered a period of extra-heavy dusting preserved in the sea mud. It began 36 million years ago and lasted 2.5 million years. During that same period, something blasted spectacular craters into what are now Chesapeake Bay and Popagai, Siberia. Based on the dust that accompanied the craters, Farley and friends blame comets.

An asteroid collision, Farley explains, would indeed generate a burst of dust that would spiral past Earth. But because most asteroids are securely gravity-trapped inside their asteroid belt, it's unlikely that big chunks of broken asteroid would hurtle Earthward along with the dust. More likely, he says, they would continue to circulate, causing more dusty accidents within the belt. So, theoretically, a big asteroid collision would generate a sudden bump in the dust rate, followed by a long string of additional bumps, lasting many millions of years.

Comets, conversely, travel *with* their dust when they dive into the inner solar system. If a swarm of comets were somehow dislodged from their normal orbit in the distant Kuiper belt or the very distant Oort cloud, there ought to be a sudden rise in dust as they race around the Sun. And in fact Farley's dust data peak at the same time that the mysterious "impactors" slammed into the ground. Then the rain of dust into the oceans fades away rather quickly, over 2.5 million years—about the time it might take for the Sun to evaporate the orbiting comets totally.

It's not unreasonable to suspect that Earth would shiver if it found itself orbiting inside a thick disk of dust that bounced sunlight away from the planet. Recall that one of dust's great powers lies in its combined surface area. In his Seattle office Don Brownlee produces a homespun illustration: "If you put a bag of sticks on the Space Needle, you couldn't see it," he proposes. "But if you burn those sticks, the smoke is very easy to see." Likewise, a comet's nucleus—the actual dust-and-ice ball—is hard to see, because its surface area is fairly small. But scatter a little of that comet's dust and you produce a huge amount of surface area. The dust that forms a comet's tail is just the tiniest fraction of the comet itself. But it is the tail that reflects sunlight and causes a comet to glow in the night sky. If the Earth were to find itself embedded in a thick cloud of this reflective dust, it might well shudder with cold.

Whether the solar system is enduring a dust blizzard or a drought, the Earth will catch its share of the spiraling specks. If a bit of space dust passes within about sixty miles of the planet, far above where airplanes fly but well below where the space shuttles orbit, its flight will be slowed by the sticky atmosphere. In just a few seconds the speed of a trapped speck will slow from tens of thousands of miles an hour, as it sustains a trillion tiny collisions with gas atoms. After plowing through just a mile or two of atmosphere, the dust will stall.

For weeks it will be sloshed around in the atmosphere, carried up and down, driven east and west with its gassy captors. It's not alone. Trapped in a cubic inch of the Earth's atmosphere are a million times more pieces of cosmic dust than there are in a cubic inch of outer space.

This exotic dust may perform an intriguing trick before it heads down to perch on your windowsill. Noctilucent clouds are terrifically high-hanging clouds that seem to glow at night. Unlike shifty aurora borealis, noctilucent clouds are stationary. Seen in summer, most often in the colder latitudes, they're a source of puzzlement to meteorologists. That's because water vapor usually cannot condense to form a cloud without the aid of dust. But noctilucent clouds form much too high in the atmosphere—fifty miles up—for Earthly dust to climb. So one theory gives the credit for these strange clouds to newly entrapped space dust.

From this lofty altitude a trapped dust grain will slowly slip down between gas atoms. About one month after encountering the Earth's atmosphere, it will gently add its weight to the planet.

Not every bit of space dust has a safe trip down. When a large grain crashes into the atmosphere, it heats up and melts. A shooting star is actually a shooting sand grain or a shooting pebble, whose mortal conflict with the atmosphere reduces it to a streak of light. Meteor showers, such as the November Leonids, are entire herds of sand-size discards from a comet: each time the Earth's orbit crosses a dusty comet trail, a little more of the grit is swept into our atmosphere and reduced to vapor. It may actually be this "meteor smoke," the vapor of sand-size or even house-size meteorites, that helps noctilucent clouds to condense.

Smaller is safer, then, for incoming space dust whose spiral carries it into the Earth's path. The size of dust grains that most frequently survive the atmosphere is about twice the width of a hair. Like miniature marbles, these "cosmic spherules" are the remains of particles that were probably twice as fat before they struck the atmosphere and melted. Though woefully diminished, they do retain some of their mass when they cool and harden. But the best-preserved space dust is the stuff no wider than a tenth of a hair.

"The magic quality of these particles is that they slow down at a very high altitude," says Brownlee. "And the density of the air ninety or a hundred kilometers up is low enough so that it doesn't shake apart the grain."

To their chagrin, Brownlee and his fellow dust hunters cannot catch all these tiny particles before they hit the ground. Nor can they induce them to settle in a clean place. And to find out exactly what sort of stardust we're made from, a serious dust scientist needs surefire dust—lots of it.

At the South Pole the Amundsen-Scott Research Station draws its drinking water from a teardrop-shaped cavity melted in the two-mile-thick Antarctic ice sheet. Through a small hole at the surface, water is pumped up from the cavity, heated, and then returned to the well to melt more ice. As the warm water eats through the ancient layers of snow, it also frees any dust that was buried along with the snowflakes.

These days polar researchers are drinking water that fell as snow during the Crusades. Hundreds of years' worth of dust has piled up on the well floor. And because only a small amount of windblown Earth dusts fly clear to Antarctica, the well is relatively rich in space dust. So when dust huntress Susan Taylor, a former student of Don Brownlee's, decided to mine the well in 1995, she was looking in the right place. But scooping them up was hardly a matter of throwing a bucket down the hole. "We had to make sure every single thing we put down there was drinking-water safe," she recalls. "It was a great adventure: high stress, not enough pay."

Taylor, a Medusa-haired and Birkenstock-shod research scientist at the Army's Cold Regions Research and Engineering Laboratory in Hanover, New Hampshire, built a very clean vacuum cleaner on the end of a string. Higher on the string she hung a camera and a light, which allowed her to watch the vacuum cleaner on a video screen and steer it from one little drift of dust to the next. She drilled a second hole into the well, so that if her gear got stuck, it wouldn't cut off access to drinking water. She lowered the vacuum cleaner 350 feet into the well. And then she hit the "vacuum" switch.

"We found *tons* of them!" she reports. "Well, not tons. Less than a gram, actually. But thousands of particles. And these little guys? You can *see* them. And they're beautiful!"

Taylor's photograph of the glamorous little guys made the cover of the science journal *Nature*. They look like black and red-brown droplets of glass. A few, Taylor says, even retain tiny tails, formed as their melted molecules streamed off them in flight. They are large—two thousand times the diameter of the primordial space-dust grains from which they grew. And that's after losing as much as 90 percent of their bulk as they whistled down through the atmosphere.

Because Taylor knew how much ice had melted in the well and how many years that ice had accumulated, she could back-calculate the rate at which space dust falls on the South Pole. She divided her total harvest of dust by years. She then divided that annual harvest by the square footage covered by the well. She was able to estimate that these tiny marbles make up the lion's share of the 40,000 tons of space dust that falls to Earth. And her detailed analysis of the dust is yet to come.

A less laborious dust-collecting experiment looks a lot like a kiddie wading pool sitting atop the North Mudd building at Caltech. That's because it *is* a kiddie pool, says Ken Farley, space-dust hunter and geochemist. Dust falls out of the sky into the plastic pool full of water. A pump circulates the water past a magnet, to which the iron-rich space dust sticks. And although a small amount of Earth dust sticks to the magnet as well, Farley is able to determine what percentage of the total accumulation is space dust.

Each week he analyzes the catch to see if space dust comes to Earth at predictable times of the year. Preliminary results do indicate bumper crops in the summer and winter, with more meager dustings in spring and fall. But his dust haul may say more about how Earthly wind and weather steers space dust to Los Angeles than it says about dust seasons in the solar system.

The kiddie-pool system is so effective that Farley has established an international network. There is a collector now on a Hawaiian mountaintop and another at Oxford University in England—although Farley admits he's updated the technology. "These things sit in the sun forever, and the plastic pool really takes a beating," he says. "So we're using surplus satellite TV dishes." The original pool retains a place of honor atop North Mudd, however.

Incidentally, the pool program has generated nice data about the rate at which regular old *Earth* dust falls on Caltech. Every three months Farley cleans about five grams—fifteen pulverized aspirins' worth—of nonmagnetic dust from the pool bottom.

The deep ocean is another good place to go looking for dust—if you have the stomach for lurching around on the waves while a drill bores a core from the deep sediments. Close to the continents the ocean floor is littered with pebbles, sand, skeletons of critters, and other big junk. But such heavy garbage can ride the currents just briefly before settling. This natural sorting system means that only the finest grains of the eroded Earth can travel far out to sea before they sink. So if you find a large particle in deep-sea muck, it's a pretty good bet it's not Earthen. It probably fell straight out of the sky.

"Less than one part in a million of deep-sea sediment is interplanetary dust," Don Brownlee says. "But for grains between a hundred and two hundred microns [one to two hairs] that are spherical and magnetic, *half* are interplanetary."

This falling dust doesn't sink immediately. The grains are too small to plummet to the sea floor. To get down, they need to put on some weight.

"I understand that all these particles have an inglorious end," Brownlee says with a little smile. "These four-plus-billion-year-old particles end up as fish fecal pellets."

Rather than waiting for dust to fall in a kiddie pool or teasing it from fish dung, Brownlee prefers dust that is fresh-caught, twelve miles up in the atmosphere. On a crowded shelf in his office sits one of his favorite ways to capture it. Smaller than a deck of playing cards and made from satiny aluminum, this little trap contains a piece of fancy fly tape called a "flag."

"I never open these unless we're in the clean room, wearing full clean-garments," he apologizes, fiddling with the box. "We put a tremendous amount of work into these particles, and we don't want to lose them."

Since 1974, NASA has been mounting sticky flags on its highest-flying planes and trolling the skies for space dust. To prepare their bodies for the low air pressure twelve miles up, the pilots have to "prebreathe" oxygen for an hour before flying, and they must wear big, bulky suits. So it should come as no surprise that NASA does not dedicate entire six-hour flights to gathering dust. The airplanes are used mainly for studying the Earth and its atmosphere. If, however, a flight is cruising at the correct altitude and if the pilot isn't overburdened with other chores, then the dust flag might fly. And fly. And fly.

"If you put the collector up there for half an hour, you'd be very discouraged," Brownlee warns. "You get about one particle an hour." And so the flag may fly on ten trips before it is taken off the plane.

The flag method delivers a better percentage of space dust than, say, a wading pool. But it's still imperfect. When the airplanes fly out of desert airfields, they carry their own little skin of dust and pollen and other terrestrial stuff into the sky, and they shed this dust, as well as microscopic flecks of airplane paint, onto the flag. The flag also catches particles of rocket fuel, beads of sulfur blasted upward by volcanoes, mold spores, pollen, and other bits of windblown biology.

"It's an interesting question," Brownlee says with a typical mix of bafflement and delight. "How high does life go?"

To avoid such distractions, and to catch dust that is absolutely guaranteed to be unaltered by the Earth's atmosphere, the obvious solution is to go even higher. In February 1999, NASA launched *Stardust*, a comet-mugging spacecraft of Brownlee's conception. As it roams the inner solar system, *Stardust* is trapping dust grains in pads of aerogel, a fine-grained foam. In January 2004, *Stardust* will go to the horse's mouth, approaching a hurtling comet, Wild 2. A camera on *Stardust* will feed Earthlings live footage as the spacecraft takes a bite out of Wild 2's coma, the dense haze of gas and dust that surrounds a comet's icy nucleus. Then *Stardust* will spit out a dust-filled capsule, which will plop down in Utah's Great Salt Desert two years later.

Because space dust holds the key to our cosmic origins, it is suddenly a hot scientific commodity. So *Stardust* is just one in a flock of dust-hunting spacecraft heading into the heavens. A spacecraft called *ARGOS,* for instance, carries an instrument named SPADUS, which will orbit Earth until the year 2002, measuring the speed and concentration of dust grains that swarm around our planet. *MUSES-C,* a Japanese-built asteroid chaser, will approach the asteroid Nereus in the year 2003. Hovering, it will shoot a bullet into the asteroid and collect the resulting dust in a hatbox-size capsule, which it will pitch back to Earth in 2005. That same year NASA's Deep Impact mission will do similar violence to a comet nucleus: On July 4 a hovering spacecraft will fire a mock meteor into the comet Tempel 1 at 22,000 miles an hour, then analyze the resulting dust with onboard instruments.

From custom-built, waterproof vacuum cleaners to space-roving comet muggers, scientists are perfecting their methods for bagging dust. But challenging as it may be to catch the dust, that's the actually easy part. It is only through the trialsome handling and analysis of these microscopic gossips that our starry past can emerge.

Caring for space dust is a fine art. First, where to put it so it won't get lost? Our nation has a dust library—and a dust librarian. NASA's Curator for Cosmic Dust, Mike Zolensky, oversees a collection of about 100,000 specks at the Johnson Space Center in Houston.

When a dust-collecting flag is taken off a NASA plane, it gets a cursory look and a quick description. The airplane's itinerary might be noted, or a brief description of the dust bits themselves. A typical note might read, "Alaska, except for flight home." Or "Background of clear and some black spheres. Some contamination." If there is time to "pick" and mount some particles, they'll be lifted out of the silicon oil that coats the flag and mounted on tiny plastic plates. But usually that job is left to the dust borrowers.

"There isn't enough money in the world to catalog all those particles," Zolensky says. "We catalog maybe three hundred a year. We routinely put away half of the flags untouched."

This hard-caught dust sits in nitrogen-filled cabinets, to prevent oxygen from rusting the dust's iron content. A catalog, describing Zolensky's two hundred flags and six thousand individual particles, circulates to laboratories worldwide. A scientist might request a flag and then pick and mount only the particles she wants, leaving the bits of desert dust, pollen, and airplane paint on the flag.

Brownlee learned the hard way how to pick dust. The early days of handling space dust, he recalls, were fraught with angst. In the 1980s one important mote simply got lost. It is still a surpassingly delicate task.

"People should know that dust is totally different from the rest of the world," he says, with uncommon seriousness. "It's *incredibly* affected by electrostatic forces. A five-micron or ten-micron particle was a little black dot under a regular microscope. You'd just get it on the pickup needle and it would be gone. They'd just jump off. We finally figured out we needed high humidity in the lab, and ion generators."

Brownlee reaches into a file and pulls out a photograph of one hundred picked-and-mounted particles. They look like a hundred pieces of gray popcorn, grapes, and rocks, arranged in a grid of white squares. Each one was transferred to its white square on the tip of a needle, under a microscope. The entire hundred-speck grid, greatly enlarged in this photograph, is half as wide as a grain of rice. And what special glue holds the precious popcorn to its grid?

"We don't glue 'em," Brownlee says happily. "Think about how dust sticks to a wall. Small particles don't come off. You almost have to pry them off. They're *dust!*"

Having learned to play by dust's exacting rules, scientists can now begin to analyze the ancestry of the individual specks—and the ancient grains within the specks.

Once dust is captured and firmly under thumb, a scientist can give it a grilling about our astral heritage. One of the first tasks is to determine whether the dust was shed by an asteroid or a comet, or whether it is virginal stardust, never caught up in a bigger body.

Brownlee pushes another giant photograph of collected space dust across his desk. Mainly these grains resemble sparkly black coconuts. Among them is a yellow item that brings to mind a microscopic wad of chewing gum. Brownlee cocks an eyebrow at it. "That? It could be from the lab. The airplane. A volcano," he says with a shrug. "It's easier to identify the extraterrestrial ones."

But it's no picnic determining their past. Brownlee is the first to admit that maternity tests for space dust leave plenty of room for speculation. "All these particles are orphans," he says. "You can prove they're extraterrestrial. You can *guess* where they came from. But there are millions of comets and trillions of asteroids." Not to mention countless specks that have roamed free since the birth of the solar system.

For now, nailing down the parentage of a given dust speck is a matter of piling up circumstantial evidence. You might start with the speck's appearance. A piece of asteroid dust may well look like a chip of solid rock. When dust is rolled up into a big body like an asteroid, heat, gravity, and sometimes even the action of water pack the dust grains into dense rock and metal. But on the other hand, if your speck resembles a piece of gray popcorn, riddled with air pockets that once held ice, it's probably cometary. The dust stored in comets never underwent the torment of heating and compression. The original grains are just loosely glued together with ice.

If this simple diagnosis fails to convince you of the speck's origin, you might try a new line of interrogation: How fast was the speck flying through space when it plowed into the Earth's atmosphere?

Comets take their dive around the Sun at high speeds and unorthodox angles, and the dust they shed typically approaches the Earth at upward of 45,000 miles per hour. (Tempel-Tuttle, whose grit gives us the Leonid meteor showers, actually orbits nearly head-on to Earth, and its discarded dust strikes our atmosphere at a blistering 150,000 miles per hour.) Asteroids, however, orbit at roughly the same speed as the Earth, and in the same direction. So when a piece of dust whacked off an asteroid reaches our neighborhood, the Earth's gravity might accelerate it to a mere 30,000 miles an hour. Virginal stardust, specks that roam the galaxy alone, are expected to travel fastest of all. But scientists suspect that only one mote in a thousand fits that description.

Exceptions to the rule abound, but generally, if a piece of asteroid dust, a piece of comet dust, and a piece of virginal stardust all hit the atmosphere at the same angle, the speeding stardust will get the hottest, and the dawdling asteroid dust will remain coolest.

To determine the speed of your speck, you'll need to know how hot it became when it hit the atmosphere. Focus your microscope on the speck until you can see the slim tracks scratched in its surface by solar flares, bits of abused atoms hurled into space by the Sun. If you find none, you might conclude that your speck was traveling so fast that its collision with the atmosphere heated it to a thousand degrees and melted over the tracks. You might also pop your dust into an oven, heat it gradually, and note the temperature at which various chemical components vaporize. If your speck struck Earth's atmosphere with such force that it warmed to, say, 900 degrees, then you'll find that no vapors emerge until you heat it to 901. With the dust's entry temperature established, an accomplished dust scholar can calculate its entry speed.

Now, if you've kept track of the ingredients you vaporized off your speck, you can move to the next question: What is the chemical pedigree of this dust?

Asteroids are grouped into distinct, chemical "families." Likewise, meteorites, most of which are probably asteroid chunks, are divided into families. And dust can also be sorted by its geology. The mineral content varies widely from speck to speck. Furthermore, some grains contain lots of water—much more, proportionally, than does Earth. Others contain thousands of times more carbon than appears in the Earthly mix. If your dust's composition is similar to the chemistry of a particular meteorite, you might surmise that your dust broke off an asteroid from the same region of the solar system. Or if a comparison of your dust to the Earth's average makeup shows huge differences, you might guess that your grain grew in a very different part of the young solar system.

The next trick is to link your piece of dust to a *specific* comet or asteroid. Since we can't trap those big bodies and analyze their chemistry, having a piece of their dust is the next-best thing. Your minuscule speck would speak volumes about the temperature, chemistry, and other conditions that influenced its old realm in our ancestral dust disk.

Does the probability of matching your dust speck to a comet or asteroid strike you as slim? You're in good company. Even the top dust sleuths rarely manage to forge a link between a speck of settling space dust and the heavenly body that shed it. But it can be done.

In June and July of 1991 a NASA airplane caught a few porous, fragile motes. A couple of dust investigators discovered that these particular motes were curiously poor in helium. In space the solar wind carries a distinct helium isotope. When a screaming atom of this helium runs into a piece of space dust, it becomes embedded. Then, should a scientist catch this space dust and heat it to about 2,500 degrees, the special helium isotope comes boiling out. So when the dust sleuths found such a small amount of helium in their dust, they knew that these specks had been exposed to the solar wind for only a few years. Now, what sort of celestial body would be most likely to release its dust so close to Earth that it would be swept up after such a short time? Because asteroid dust takes thousands of years to spiral into the Earth's neighborhood, the riddlers concluded that this dust must have been dropped by a passing comet—and quite recently.

The next clue they found was that this dust hadn't gotten very hot on its trip through the Earth's atmosphere. That meant it had entered slowly. There-fore, its parent comet must have crossed the Earth's orbit slowly.

After considering the list of candidates, the investigators found just one comet that could have dropped the dust: Schwassmann-Wachmann 3, which crossed our orbit in May 1991. By carefully interrogating tiny grains of dust, these sleuths had secured a sample of a distant, hurtling comet.

Even after scientists link one grain of dust to a specific birthplace in our ances-tral dust disk, more work remains. They must then look deep into the tiny speck itself. Each of the 100,000 grains that make up one dust speck is, ulti-mately, primordial. But most of these bits of stardust have been melted by heat or water or crushed into new forms by gravity. They have reacted with the chemicals around them, until they no longer bear the mark of their birth. Only a few micrograins survive to drop dazzling hints about the long-gone stars that produced them.

Silicon carbide, graphite, sapphire, and diamond are stardusts so durable that some of them have survived all the churning and heating that the Sun's violent birth could dish out. While weaker stardusts merged and melted and re-formed, these hardheaded specks endured. So when scientists find these micrograins inside a meteorite or a piece of space dust, they view them as little ambassadors from long-dead stars.

It was not always so. The first diamonds to be discovered in a meteorite were teased from a mysterious rock that plunged to the ground in Siberia over a century ago. This meteorite fell to Earth as one of a pair. One piece was deliv-ered to scientists. Its twin met a less exalted fate: it was reportedly eaten by farmers, presumably after undergoing some pulverization. And still, a century after their discovery, space diamonds are holding their secrets tightly. Because they're so devilishly hard to detect, scientists can't even get direct proof that di-amonds reside in dust grains.

"I'm *sure* there are diamonds in the dust," says Don Brownlee. "Microdia-monds are abundant in primitive meteorites, and surely they exist in cosmic dust. But no one's actually *seen* them yet because they're difficult to separate from the background. So we're dissolving away the mineral matter and just looking at the carbon. We *think* we see them."

Recently, Danish astrophysicist Anja Andersen calculated that a whopping 3 percent of all the carbon that gathered to form our solar system may have been in the form of microdiamonds. That is the percentage that turns up inside pristine meteorites, Andersen says. And that may be the tip of the diamond iceberg. "Since much of the original stardust was remelted during the solar-system formation," she notes, "it is reasonable to expect that there was *more* diamond present at the start."

To get a close-up view of a space diamond you might peer into any handy piece of diamond jewelry. Diamonds mined on Earth may incorporate a tiny space gem, says University of Massachusetts geoscientist Stephen Haggerty. Work with laboratory-grown diamonds has shown that diamonds grow easiest and fastest if they are given a "seed" to crystallize upon. And in the Earth's crust, "diamond seeds" from space may be a dime a dozen. After all, the sort of meteorite that carries microdiamonds delivers *huge* concentrations of them to Earth—up to 1,600 times more diamond, pound for pound, than is found in the best diamond mines. So it's entirely reasonable to assume that some cosmic diamonds grew quietly underground into the macrogems that now flash rainbows from human fingers and ears.

If space diamonds are closemouthed and secretive, then GEMs—"glass with embedded metal and sulfides"—are more forthcoming. These are smoke-small, roundish dusts, made mainly of glasslike silica but studded with glittering crystals of metal. Under an electron microscope a GEM looks like a gray pebble dressed in flashy sequins. In 1999, Brownlee and teammates announced that GEMs qualify as another antique micrograin hidden inside space dust, and that they have a story to tell.

To get a good look at GEMs, Brownlee's team sliced microscopic specks of space dust into thin sections. Then, using an exquisitely fine-sighted instrument at the Brookhaven National Laboratory, they analyzed the chemistry of the glittery micropebbles scattered throughout the slices. When the analysis was complete, another link had been forged to our distant past: the chemical signature of the GEMs matches the chemical signature of distant clouds of cosmic dust. Anything the team now learns about captured GEMs will tell them more about how dust developed inside our own ancestral dust cloud.

The GEMs have already produced a bold clue about the sort of special ingredients that may have been rolled into our planet. The micropebbles are coated in thick layers of intriguing, carbon-rich material. These black coats, the team concluded, probably formed in the cold, dusty cloud even before the Sun and planets took shape.

Dust scientists are surprisingly close to the goal of describing the group of long-gone stars whose dusts hatched our solar system. Although those "ancestars" blinked out billions of years ago, their voices whisper to scientists from the minute grains of space dust they left behind. To reconstruct those stars' identities would be a monumental step in a long, slow journey.

Scientists started down this road about two centuries ago. A spectacular shower of shooting stars in 1833 inspired a Yale astronomer to suggest that the cause might be thousands of exceedingly tiny meteorites. He even compared them to the crumbs that make up comet tails. Soon other astronomers were pairing up the paths of specific comets with the showers of burning grit.

The notion that super-small meteorites were actually landing on the Earth also circulated quietly at this time. Cosmic dust was been proposed as the agent of red sunsets and the source of metal that faintly tainted raindrops. Then, in 1872, the HMS *Challenger,* a two-hundred-foot sailing marine laboratory, left England to sample ocean sediments around the world. The onboard naturalist, Sir John Murray, used a magnet to lure tiny, shiny spherules out of the ocean's remotest muds. Comparing their chemistry to that of meteorites, he declared they were indeed wee cousins.

The prospect of a gritty cloud swarming around the planet took on new urgency when people began sending delicate machinery into space.

"In the beginning days of space flight they were worried that spacecraft wouldn't survive the dust," Don Brownlee says with gentle amusement. "People thought the flux was a million times more than it is. They thought satellites would last only a year." Starting in the 1950s, spacecraft were fitted with shields to fend off what was imagined to be a howling gale of abrasive dust.

But now that scientists are sending their machines out for the express purpose of gathering dust, we've turned a corner: Cosmic dust is no longer a threat. It's a promise.

Can we hope that dust will eventually let us identify the exact parents of our solar system? It may show us the *types* of stars that gave rise to our world, if not the culpable individuals, says Larry Nittler, an astrophysicist at Goddard Space Flight Center. Nittler dissolves samples of meteorite in strong acid. This frees the tough, ancient dust grains so that he can analyze their chemistry. By comparing the ratio of the isotopes, carbon 12 and carbon 13, Nittler can narrow down the type of star that made the dust. That's because each class of stars, when they are grouped according to their size and age, produces dust with a distinct mix of carbon isotopes. The isotopes of aluminum oxide—sapphire, to a jeweler—in a speck of meteorite also come in distinct ratios.

But despite the finesse of Nittler's measurements, the exact number of stars that contributed dust to our birth cloud still eludes scientists. "It's much more than one," Nittler says, laughing. "Based on several lines of evidence, most people think it's between ten and a hundred. The important thing is, it's a lot."

The diversity of sapphires alone, Nittler says, suggests we owe our respects to about thirty individual dead stars. But addressing the Mother's Day cards will be difficult. "It's impossible to tie a grain to a certain star," Nittler says. "Those stars were already dead when the solar system formed billions of years ago."

But our ancestars left us tokens of their brilliant lives. Like a diamond ring that is passed from grandmother to mother to daughter, jewel-studded space dust drifts down onto Antarctic glaciers, rooftop kiddie pools, and our kitchen counters. And locked in this dust are tiny reminders that we, along with every other bit of life on Earth, are part of a very long and extraordinary story.

4

THE (DEADLY) DUST OF DESERTS

Something horrible befell Big Mama.

Big Mama was a dinosaur. Technically an *Oviraptor,* she was ostrich-sized and roughly ostrich-shaped, if you could overlook the wide, turtlish beak and the curved claws as long as human fingers. When explorers encountered the first *Oviraptor* skeleton in the orange sandstone of the Gobi Desert in 1923, they grievously mistook this animal's identity. Because the skeleton sprawled near a circular nest of fossil eggs, the species was branded, in Latin, "egg thief." It wasn't until 1993 that additional fossils redeemed the animal's reputation. One fossil was an egg of the same type, inside of which curled an embryonic *Oviraptor.* The second fossil was an adult *Oviraptor* hunkered so squarely and competently over yet another batch of eggs that the mistake was clear. Not *Oviraptor,* but Oviprotector. Although paleontologists can't be sure of the gender, this fossil was nicknamed "Big Mama."

This poster child for good parenting had lain buried for roughly 75 million years, with its future reduced to ovals of stone beneath it. Finally carved free and picked clean of ancient sand grains, the creature's bones and eggs did argue eloquently that paleontologists underestimate the care dinosaurs invested in their progeny.

But the surfacing *Oviraptor* raised a fresh question: What buries a dinosaur so fast and so deep that she never has a chance to hop off her nest and run?

"In a sense it's an ancient murder mystery," drawls Lowell Dingus, the big-hearted geologist who redesigned the famous dinosaur halls at New York's American Museum of Natural History, and who has spent many a July shuffling through the hundred-degree-plus heat of the Mongolian Gobi.

Dust, of course, is a suspect. But it was a complicated crime.

For one thing, the faithful *Oviraptor* wasn't alone. Over eons, many other kinds of dinosaurs, reptiles, mammals, and birds were slain here, as abruptly as the nesting *Oviraptor.* The victims ranged from tanklike ankylosaurs to turtles to tiny, primitive mammals. And now, in rich pockets all over the Gobi, which sprawls on both sides of the Mongolia-China border, the carelessly packed sandstones are crumbling and letting go of these bones. Vertebrae. Teeth. Rib fragments. "It doesn't *look,*" Dingus says with a wry grin, "like they all got arthritis at once."

Whatever method dust employed to massacre these animals, it did paleontologists a mammoth favor. As the *Oviraptor* illustrates, the Gobi skeletons actually preserve dinosaur *behavior.* And in the dinosaur business, that is not the norm. Usually, a dead dinosaur would have been gnawed upon, torn apart, and scattered over an acre of land. Then acids in the soil would have gradually reduced the scattered bones to a subtler version of the rubbery joke that results from soaking a chicken's wishbone in a glass of vinegar. Even if a dead dinosaur were lucky enough to fall into a protective river, the law of averages would demand that his bones be scattered as his corpse rotted. To make a good fossil, a bone must be buried quickly. The odds are poor. To make a good *skeleton,* the entire animal must be buried quickly. Those odds are extravagantly poor.

In the Gobi, though, it is common to discover complete dinosaur skeletons frozen in lifelike poses. Also puzzling, a great number of extraordinarily *tiny* bones have been preserved. Some of the glossy, fossil eggshells hold the toy-size white bones of embryonic dinosaurs. Even our own mammalian forerunners, shrewish creatures with leg bones like pencil leads and vertebrae like grains of rice, are here. Fossil hunters scour the ground for half-inch balls of sandstone that when turned in the hand reveal the white skull of a mammal or a tiny lizard. The leaf-thin bone is packed with sand-turned-to-stone, which glows pink in the eye sockets and between the tiny, toothy jaws. A month of hunting in the Gobi can net fifty skulls. No place in the world compares. Why?

Pestered by that tickling question, in 1996 Lowell Dingus enlisted the Doctor of Sand, David Loope, a tall, taciturn, sand-dune expert from the University of Nebraska at Lincoln. Loope formulated a theory involving dust. And in 1997, as part of a dinosaur-hunting research team of eighteen people, the two geologists ventured back into the Gobi to sift out the proof.

It wasn't a quick trip. There were no gas stations, ice machines, pay phones, or any other conveniences along the trail that headed south from Ulan Bator. Mainly, there were sparse wild chives, gravel, camels, and blinding quantities of sunlight and dust. And so food, tools, tents, water tanks, spare tires, jugs of transmission oil, beer, and quantities of toilet paper were stacked in two trucks. During frequent breakdowns Dingus smoked cigarettes in the back-seat of a Russian four-by-four, a green felt hat crunched over his dark glasses. The Doctor of Sand napped, a red tractor cap shading his face. Camel ticks, sensing opportunity, came trundling from the skimpy shade of prickly silver-green shrubs. The better part of a week passed before the convoy paused on a hilltop overlooking a wide gully studded with intriguing red buttes.

When the dinosaur hunters stepped out of the trucks, flecks of dinosaur eggshell snapped under their boots The ivory-colored shards were everywhere. And every footstep raised a little puff of dust.

Mongolia, like every other patch of Earth, is made of melted dust. As our molten ball of space dust cooled, the lighter minerals floated to the surface. Driven by currents in the still-molten rock below, pieces of this cooling crust migrated slowly around the globe, carried on plates of sea floor. The young continents blindly crashed into each other and retreated. As liquid rock rose and cooled, the continents grew.

The stretch of crust we call Asia began rising out of the sea like a reptile, headed by Siberia, just half a billion years ago. Mongolia, this reptile's shoulders, followed slowly, with the sea periodically sloshing over its skin. There are fossilized corals in the south and west of Mongolia, now many hundreds of miles from the nearest ocean. And the eroding mountains of the Gobi hold limestone deposits, probably composed of the skeletons from ancient sea creatures.

The early dinosaur era in Mongolia, long before the *Oviraptor* evolved, may have been quasi-tropical, and dustless. Warm winds, sodden with water vapor from the sea, swept monsoons north across the land. But south of the reptile's belly, arc after arc of volcanic islands were thrusting themselves out of the ocean, only to be shipped north and rammed against the lower edge of young Mongolia. China was being made. One after another, fresh stretches of crust were plastered onto Asia, setting Mongolia ashudder and crumpling its rocks into tall, parallel ridges. And in China new mountains were growing, too, and eventually they rose so high that the sodden monsoon clouds traveling north were forced to toil up and over them. As each cloud ascended, it would cool, and its water vapor would condense. Rain would pour out of the cloud as it climbed the south flank. If any vapor remained, the diminished cloud would float on over the range, to throw only mocking shadows on the thirsty corrugations of southern Mongolia.

A desert was born, and the dust began to rise.

· · ·

At Ukhaa Tolgod in the modern Gobi, Dingus, Loope, and the research team erect a herd of yellow tents on a rolling, gray-pink plateau. By day, fossil hunters will be blasted with clouds of hot dust and sand, and on windy evenings they'll eat quickly to protect their food from the flying grit.

A few steps from the camp's fire pit a trail descends to the valley floor, which is about a mile wide and generously spattered with fossil fragments. More white bones protrude from the cliffs of yellow and orange sandstone that enclose the valley. One July morning Loope and Dingus tromp down into the basin and head toward the right-hand wall. At eight o'clock the sky seems violently hot. The microscopic dust that hangs in the air transforms the sunlight into a relentless glare. The pebble-paved ground ripples with heat. On the horizon a few slow camels shamble among tortured tufts of shrubbery.

Loope is only occasionally moved to speech. Dingus practices a similar linguistic frugality. The boisterous fossil hunters have nicknamed them the "Geo Guys."

The two men scramble doggedly up the pile of cliff crumbs at the base of the wall. On close inspection this part of the cliff appears to be divided into layers that lie on a slant. Erosion of the valley walls, Loope notes, has sliced open an ancient sand dune, whose migration was halted in the distant past. Each faint layer in this cross-section marks a time when the dune stood still and its slanted surface hardened into a little skin. Then another period of howling winds would blow a few inches of fresh sand over the dune. To the Geo Guys, such sandstones are "stratified."

Scrambling for traction, Loope rounds a corner and bends close to the layered cliff face. He points to where one pale layer makes a series of hand-size dips into the darker layer below. He squints to disguise his excitement.

"I think they're dinosaur tracks in cross-section," he says.

Dinosaur tracks are a lot easier to spot in ancient mud than old sand. But in 1984 the Doctor of Sand discovered fossilized bison tracks in the Nebraska Sand Hills. And he was the first person to find dinosaur tracks in this difficult desert.

These tracks are an indication of how dusty the Gobi was 75 million years ago. At least some of the dunes, some of the time, were naked and footloose. But Dingus says there is also evidence of small ponds and plants preserved in these old sandstones. And that means the Gobi back then was probably a bit damper and greener than the Gobi now—but not much.

How would such hulking creatures find enough to eat in such a bleached and wind-battered landscape? Well, the big carnivores could have eaten the loyal *Oviraptors* and the prolific *Protoceratops* dinosaurs, whose pig-size, frill-headed fossils are common enough at Ukhaa Tolgod to invite kicking. But what sustained *them?*

Take a closer look, Dingus urges. Consider how many animals make a living in the modern Gobi. Awkward gangs of camels. Small clots of domestic sheep. Gazillions of gazelles. There is also the occasional eagle, wolf, and hedgehog. Even a few impressive units of *Homo sapiens* have erected their felt yurts that sit like white dots in the immense desert. And this elaborate, modern food web is anchored with, as far the casual observer can discern, a few thorn bushes and wild chives. Clearly, a dusty landscape is not a lifeless landscape.

On the other hand, perhaps the tracks in the dune date to an especially arid time, and the scattered dinosaur nests date to a wetter period. All deserts have minor mood swings, from dusty to damp and back again. Big climate cycles produce even bigger changes in deserts. When ice-age glaciers locked up huge amounts of the planet's water 18,000 years ago, naked sand covered half of the land in a 4,000-mile-wide belt around the equator. These days just 10 percent of that belt is sandy. And, accordingly, the cloud of desert dust rising to circle the planet is now thinner.

In fact, little mood swings in the Gobi may have abetted dust in its mass murder at Ukhaa Tolgod.

The sandy grave of the *Oviraptor* proves that this neighborhood was dusty then. It's indisputable that the Gobi was busily disintegrating. But as Big Mama brooded her eggs at Ukhaa Tolgod, she was not alarmed. The making of desert dust is a subtle process. It is easy to forget that it is always under way.

The wind should have been a reminder. Even with no trees or grass to rattle, the wind in the Gobi is audible. As it streams around outcroppings of rock, it makes a low thrumming noise. It sizzles as it flings grains of sand at pebbles.

Most rocks, from mountains to deposits of limestone and even grains of sand, harbor weak spots. Since the planet's rocks solidified from molten dust—or from accumulated coral skeletons—they have been thoroughly abused. In the past 4.5 billion years, most rocks have been folded, twisted, mashed, or turned upside down, have fallen off a cliff, been ground to sand or melted to lava—and some have endured all of the above. So a rock is not so much rock-solid as it is riddled with defects and tiny fractures. A limestone cliff is an accident waiting to happen. A grain of sand is as good as shattered when the wind prepares to drop a tiny grenade upon it.

It takes a wind of at least twenty miles per hour to lift sand from the ground. As the wind roars around it, a sand grain on the desert floor will jiggle back and forth. And then it will take wing. It will sail for a short distance, climbing no higher than about six feet. And then it will plummet back to Earth. When it lands, it might strike the corner of another grain, launching a shower of tiny chips: dust is born.

Furthermore, the impact of this grenade may be the boost required to bounce the victim up and into the wind. And so a new grenade perpetuates the cycle. Windblown sand can also break flecks of dust out of pebbles and boulders. It can blast a shower of pale limestone dust off a mountainside.

Frost is a more subtle dustmaker in the Gobi, which is actually rather sand-poor compared to other deserts. It gets a modest amount of rain, and snow in the winter. This moisture seeps into the slim cracks and fractures in rock. Then, when the temperature falls, the water turns to ice. Its crystal structure pokes out its elbow, forcing the rock fracture to open wider. Whether water invades a crack in a cliff, in a boulder, or in a grain of sand, it will eventually shatter its host. If it fails on the first freeze, it will thaw in the next warm spell, seep in deeper, and refreeze in a position of superior leverage. Sooner or later a microscopic avalanche of dust will fall.

Salt employs a similar method of splitting dust off rocks. Various salts, which are common in the Earth's crust, naturally leach out of aging rock. But deserts, deprived of the flushing action of water, are especially salty. So perhaps, as the *Oviraptor* sat daydreaming on its nest, an airborne crystal of salt settled onto a pebble of limestone at its side. When night fell, the salt was moistened by a drop of dew. The salt melted and slipped into a fracture in the pebble. The next morning, when the Sun rose, the dew evaporated out of the crack. And in its new position, the salt returned to crystalline form—and rammed its elbows into the pebble. If Big Mama could have stood in that microcanyon, inside that pebble, she might have heard a tiny groan as the canyon walls were forced apart.

In the modern Gobi this is the terrible trio: wind, salt, and ice. They get special assistance from the precarious landscape in which they operate. Rocks that are being actively tormented are the easiest to erode. And the India-Asia collision that has been building the Himalayas for the past 50 million years also puts the squeeze on Mongolia, a good thousand miles north of Mount Everest. The Gobi's gray, accordion-folded mountains deliver a steady supply of falling rock to the desert basins. And there the wind, salt, and ice begin busting it down to dust.

Some rocks are naturally more dust-prone than others. Limestone and feldspar and gypsum are easily crumbled to powder. Quartz, on the other hand, is tough. Elderly sand dunes and many beaches are composed of super-tough quartz balls, worn round by abrasion. But even for this diehard mineral, dust is inevitable. A little dew is enough to infiltrate the very crystal structure of the quartz. The water traces the irregularities in the crystal and melts out one molecule at a time. Magnified, an aging sand grain looks like an aging mountain: it is rounded on top but grooved with "ravines" carved by tiny "rivers." And at the base of a "cliff" is a sprinkling of "talus," fallen boulders one-hundredth of a hair wide.

One day in the ancient Gobi a stiff breeze suddenly rattled through Ukhaa Tolgod. Bits of microtalus mixed with powdered limestone, gypsum, salt, and other dusts, swirling upward in the hot wind. And when the wind dissipated, the dust was left hanging high over the head of an *Oviraptor*.

The dust hanging over Big Mama wasn't exclusively of local origin. Erosion was steadily churning out dust all over the world. On the greener parts of the planet the dust was quickly trapped by plants and damp soils. But dust that was produced in dry parts of the planet was likelier to take to the wind. And a tiny percentage of that dust rose so high that it traveled thousands of miles before it settled. This background haze contained a little of everything the wind can scour from the surface of the Earth.

Which dusty patches of the Earth contributed the most to this high-flying background of dust millions of years years ago is a matter of guesswork. But a tour of the dusty spots on the face of the modern Earth will illustrate the variety of dusts that conspired in the ancient sky above the *Oviraptor*.

Today the master duster is the great Sahara. Most deserts are upholstered in bedrock, stones, and gravel, with a little sand thrown into the corners. But about one-fifth of the Sahara is covered with sand, and the dunes can rise one thousand feet. Imagine four Texases, buried under sand to the height of a five-story building. And imagine all that sand hopping around bombing its brethren and quietly fragmenting in the salt and sun. How much Saharan dust wandered as far as the Gobi Desert in the *Oviraptor*'s day is hard to say. But we do know that enormous gold dust clouds depart from North Africa today. Scientists are quick to note that it's a tricky business to estimate the tonnage of dust that rises from the Earth's surface. But over the years some brave souls have ventured guesses. One popular estimate says the Sahara hurls about 600 million tons of dust into the sky every year. Another estimate puts the annual cloud at a *billion* tons. At the lower rate a boxcar of Saharan dust would leave

Africa about every four seconds. Every minute, sixteen cars. Every hour, a thousand. Day after day, year after year.

China's expansive—and expanding—deserts may hurl a similar quantity of dust skyward. And the eroding farmlands of the central and western United States may throw up nearly a billion tons of soil each year, although this dust doesn't tend to travel far.

Another rich source of modern dust is *ancient* desert dust. South of the Gobi is an agricultural wonderland called the Chinese Loess Plateau. The stone-free loess (pronounced like "puss") is pure dust that has blown off the Gobi and Taklamakan Deserts over the course of millions of years. Sprinkling down at a rate of an inch every two and a half centuries, it piled up like a thin and constant snow on more than a hundred thousand square miles. That's about the size of Colorado. And if you add the other patches of loess scattered around northern China, the dust soil covers one-fifth of the entire nation. And although settling and erosion have reduced the dust blanket, it is still three hundred feet deep in many areas, a thousand feet in one spot.

Worldwide, loess covers 10 percent of all Earth's surface. Some of this ancient desert dust is safely trapped under windproof armor of vegetation. But other areas, often those abused by human beings, toss uncountable tonnage of old dust to the winds.

A glacier may seem like the inverse of a desert, but glacier dust must also have hung over the *Oviraptor*. When a deep river of ice grinds slowly through a mountain range, it scrapes off boulders the size of houses, as well as microscopic fragments called "rock flour." You can see these scourings channeled into dark lines that run the length of a glacier, like stripes on a skunk. The streams of melted ice that trickle out of glaciers are often a gorgeous shade of gray-green, due to the rock flour they carry. And when the streams deposit their load of powdered stone on an outwash plain, that dust is ready to roam. But dust science is young: scientists have no idea how much dust glaciers produce today, nor how much they ground out millions of years ago.

Dry lakes, or playas, are also generous dust donors. In the western United States alone, more than a hundred dry lakes spot the deserts. Deprived of their water source when the last ice age ended, these lakes evaporated away. From an airplane their salty skins look like pale, flat scars on the landscape. But under the salt they harbor deep deposits of fine silt washed into them by ancient rivers.

Some playas keep a tight grip on their dust. The space shuttles sometimes land on one of these, the former Rogers Lake in California's Mojave Desert. Even though traffic breaks up the protective skin, the winter rains refinish and seal the surface each year. Other lakes, with looser skins, can give up huge amounts of the fine sediment they collected during their watery lives. Today these dusts can do terrible damage.

Eighty miles north of Rogers Lake, Owens Lake enjoys a dubious distinction as the nation's single most appalling source of dust pollution. This 110-square-mile lake was intentionally drained of its water in the 1920s, to quench the growing thirst of Angelenos. Since the lake's last gasp, winds may have whipped off a quantity of dust that seems unbelievable: Estimates range from 400,000 tons to nearly 9 million tons—*each year*. Imagine emptying a five-pound bag of flour into the wind. Then imagine emptying three hundred bags every minute, year-round: that's the low estimate. The residents of the lakeside town of Keeler cohabitate unhappily with what they call "Keeler fog," a salty dust fine enough to infiltrate their houses. In 1995 Keeler's air set a record when it carried twenty-three times more dust than the national health standard allows. Only in 1998 did Los Angeles agree to bandage the vulnerable expanses of dust with gravel, water, and plants.

Bigger lakes offer even more dust to the wind. The gigantic Aral Sea, much shrunken by irrigation, hasn't managed to form a windproof crust over its sediments. One rough estimate puts the Aral Sea's dust production at a mind-numbing 150 million tons a year. Adding a modern flavor to playa dust, Aral's sailing sediments are heavily spiked with toxic pesticides.

All these mineral dusts add to the high haze of flying particles. The minerals and metals of rocks are joined by a thousand other natural dusts, from specks of sea salt to glassy microsponges of volcanic ash. And so it was in the *Oviraptor*'s era. The skin of the Earth was falling apart, and a little bit of everything was drifting overhead.

In the age of the *Oviraptor* perhaps the high haze of world-traveling dust was a bit thinner or thicker than it is today. But it was there. Even if the only dust at the wind's disposal had been a purely Mongolian powder of tormented rocks and old coral reefs, that would have kept the air plenty dusty. As regular as rain, the wind sprang up and the dust rose off the ground. The pink-orange mixture thickened the air for minutes, or even days, until the wind relented.

Was it dust storms like these that killed the dinosaurs at Ukhaa Tolgod? If the Geo Guys are right, dust storms did choke the dinosaurs—but indirectly.

To this day the short-term implications of dust storms are usually more of an inconvenience than a mortal threat. Nonetheless, the immediate impacts of a dust storm are nothing to sneeze at. They are sufficiently miserable events that desert-dwelling people of Central Asia and Arabia give special names to the winds that bring them. If you didn't know the winds spawned throat-coating dust storms, the names might sound pretty romantic: *khamsin, harmattan, haboob, afghanet, shamal.*

The physical effects of these dust-loaded winds are not limited to the application of stucco to a sunscreened cheek, nor even to the crust that forms in the nostril and ear canal. If you've walked barefoot on a summer beach, you know that sand can get much hotter than the air itself. And when lifted into the air, hot grains of sand and dust become airborne radiators, heating the air—and the people and animals—around them. What's more, the friction of blowing sand and dust creates static electricity, which is reputed to cause nasty headaches. Sand and dust storms are viciously dehydrating to both plants and animals. To compound the misery, loading up the local air with dust actually reduces the chances that rain will fall to cool things off. In addition to tormenting animals, dust storms can also shred the leaves off plants and gum up modern machinery—and may even be able to bring down an airplane.

In January 2000 dust may have driven a Kenya Airways jet into the ocean, killing all on board. The jet had flown across Lagos, Nigeria, where the airport was closed due to dusty winds, according to local news reports. The plane crashed later that night, taking off from Abidjan. Some dust experts conjectured that dust buildup may have caused turbulence inside the engines, damping their power.

On the bright side, a full-blown dust storm isn't very easy to launch. A little dust devil is easier to pull off. Like miniature tornadoes, dust devils are born when a thin layer of air lying against the hot ground bubbles up through cooler air and starts spinning. The fast-rising hot air lifts dust—and kangaroo rats—off the ground.

The kangaroo rat has proved to be a useful unit of measurement for the upward speed of a dust devil. Or so concluded one researcher in 1947, according to the scientific literature: Having noted that dust devils sometimes snatch up these luckless creatures, the investigator measured the speed at which a kangaroo rat falls when dropped from a tower. From this observation he was able to calculate that the upward speed of dust devils must be at least twenty-five miles an hour. He further observed that the kangaroo rat was angered by this employment but unhurt. Since California's Mojave Desert, as an example, can host thousands of dust devils a day, evolution may have furnished the oft-lofted kangaroo rat with a crashproof anatomy.

Although dust devils are just a few feet in diameter at the ground, they nonetheless pump plentiful dust into the air. If they find good grazing, their appetite for dust can be boundless. The scientific literature on dust devils re-counts an instance in which a dust devil parked itself over an exposed mound of dirt at a construction site and proceeded to transfer more than a cubic yard of dirt per hour from Earth to sky. After four hours of this the crew thought to park a bulldozer over the dust devil's buffet.

If full-grown dust storms are harder to start up, they're also impossible to stop. In dusty parts of the world, enormous dust storms are as inevitable as are snowstorms are in the Alps. The Alps have blizzard seasons, and deserts have their dust seasons. In Egypt, for instance, the dusty season arrives in December and tarries for five months. Springtime is dust time in Saudi Arabia. In Arizona, between May and September, drivers on the famously dusty Interstate 10 are urged to tune their radios to a channel that broadcasts alerts when a yellow-brown wall of dust rolls out of the desert. Arizona's dust storms cause forty-plus traffic accidents in an average year, plus scores of injuries and a death or two.

Although Phoenix, Arizona, is hit about fifteen times a year, it gets off easy. One of the most dust-stormed places on Earth is the Seistan Basin in eastern Iran, where in one banner year big dust storms reduced visibility to two-thirds of a mile on eighty different days. About forty dust storms per year visit Turk-menistan, Uzbekistan, and Tajikistan, which play host to the deliciously named Karakum and Kyzylkum Deserts. China's enormous Taklamakan Desert produces about thirty storms a year, as does the arid juncture of Pak-istan, Afghanistan, and Iran. These are averages, and they're vulnerable to the whims of weather and climate. The terrible drought that struck the sub-Saharan Sahel in the early 1970s caused five times the usual number of dust storms to roar through the Sudan.

Dust storms were probably a normal part of the *Oviraptor*'s life 75 mil-lion years ago. Perhaps one spring afternoon, when a newly warmed parcel of air collided violently with a cold front, a dust storm scratched across the ancient Gobi. A blizzard of sand bombed loose the dust embedded in the desert surface. Even the grains that rolled along the ground shook off bits of dust that clung to them. The sand stayed low. And the dust rose in a wall thou-sands of feet high. It billowed forward like a ground-hugging, orange thunder-head.

Had the wind been feeling its oats, it might have carried this Gobi dust high enough that it would take a week to come down—in North America. Had the wind been modest, this dust might have settled on the Chinese Loess Plateau. But the wind died away quickly. It left the dust hanging in the bright air over the Gobi.

. . .

Particularly violent sandstorms may have directly slain some Gobi dinosaurs—that is a long-standing theory. And although Loope and Dingus no longer believe that hurtling sand and dust directly massacred the Ukhaa Tolgod animals, the theory may still fit elsewhere.

A few days of slow driving north of Ukhaa Tolgod, Mongolia's famous "fighting dinosaurs" were found in the sandstone by a Russian and Polish team. When this fossil was uncovered, a meat-eating *Velociraptor* appeared to have his curving claws enmeshed in the skeleton of a piglike, hook-nosed *Protoceratops*.

David Fastovsky is a University of Rhode Island geologist who analyzes layers of sedimentary rock for clues about the ancient environment. When he went through the fighting dinosaurs' sandstone neighborhood with a fine-tooth comb, he found the distinctive layers built by windblown sand. He's convinced that sand- and dust storms can be deadly. They certainly are today.

In the Dust Bowl drought, when millions of people abandoned their homes in the face of relentless "black blizzards" of fine earth, cattle and rabbits reportedly died by the thousands, from suffocation and from eating too much dust on their greens. And there were scattered reports of people suffocating when they were caught outside. More human casualties came from "dust pneumonia," a feverish disease that was curable if it was treated quickly. And as recently as 1998, an enormous dust storm killed at least a dozen people in China's western desert region, though news reports didn't clarify whether they died in dust-related car accidents or other mishaps.

Flying sand alone can be brutal. It is a serious demonstration of Mongolian manhood to strip naked and take the flying sand on the chin. It can cause bleeding. It can cause disorientation, and death by dehydration.

"In a desert sandstorm it would be very easy for two dinosaurs to be overcome by blowing sand, then be buried by the sand," Fastovsky asserts. "It could happen very, very quickly."

During the 1997 expedition even the minor sandstorms that visited Loope and Dingus's research camp at Ukhaa Tolgod seemed to have killing on their mind. Rolling forward like a yellow mirage, the winds would thrash sand and dust out of the gravel pavement and swish it viciously among the trucks and yellow tents. On human skin the sand grains stung. They swirled into squinting eyes and flew deep into ears, registering with a *ting!* Even those tents that didn't slip their stakes to bound across the desert would afterward be coated inside with a fine pink powder. Fastovsky has been treated to comparable demonstrations of the Gobi's flighty geology.

"That convinced me that if I was a fighting dinosaur in a sandstorm, I'd be pretty unhappy," he says, chuckling.

But Loope and Dingus believe that at Ukhaa Tolgod sand had an accomplice. Toiling along the valley wall in the rising heat, the Geo Guys reach a section that launched Loope's theory.

This part of the wall is uniformly red, showing no layers at all. And on close inspection it holds a democratic hash of large and small sand grains and even pebbles, all stirred together. The paler sandstones, those that show neat layers and dinosaur tracks, are fossil-poor. But this darker, chaotic sandstone, which fills the space *between* the preserved dunes, is itself filled with bones.

And that segregation is why the sandstorm theory about the massacre of Ukhaa Tolgod's dinosaurs was starting to bother Dingus. According to that theory, Dingus explains in his leisurely manner, the simpleminded *Oviraptor* was sitting on her nest when a sandstorm came tooling along and covered her up. Energetic speculation could produce no other reasonable method for dumping tons of sand on the landscape.

"Early on, there was an argument that these sediments might be lake deposits, but that didn't seem likely," Dingus says, his brows furrowing behind his sunglasses. "Rivers? Usually rivers have enough velocity to move the dinosaur bones around. But a sandstorm: that's one of the few viable ways you could get a specimen buried rapidly without breaking the skeleton apart."

Loope, the sand expert, also had trouble with the scenario. "One of the reasons that David started questioning the sandstorm theory is it turns out a sandstorm doesn't deposit more than a couple feet of sand," Dingus says with a grin. The reason he finds this amusing is that the *Oviraptors* were *Velociraptor*-style dinosaurs that stood three feet high at the shoulder. They had long, powerful legs, and terrifying claws. An *Oviraptor* skeleton does not suggest an animal that would have lain down and died because a bit of sand fell on its nest. And these critters were frail little things compared to some of the other dinosaurs also found in the jumbled sandstones. The more the Geo Guys thought about it, the less they—and their dinosaur-digging comrades—liked the sandstorm story.

Loope taps the red wall, raising his voice against the wind. "I think these unstructured units are a sediment flow—a *sand*slide."

. . .

On a spring day about 75 million years ago, a sudden wind shot through the Gobi. It lifted the sand and threw it against an ancient deposit of limestone, or calcium carbonate. A thin river of this white dust wandered up into the sky. The fine particles danced in the air over huge dunes tinted with a sparse net of springtime foliage.

A day or so later a passing cloud released a raindrop. And as the drop plummeted through the haze of dust, it engulfed a lingering speck of limestone dust. Inside the falling drop perhaps that tiny fleck of limestone started to dissolve.

When the drop crashed into the steep face of a dune, it threw the melted limestone into an eight-hundred-foot hill of sand that had retired from the traveling life. Most of the water in this raindrop evaporated quickly and rose back into the shimmering air. But a microscopic river remained, running down between the grains. It carried with it the dissolved limestone. And when this thin river petered out about three feet below the surface of the dune, the limestone resolidified on the warm, smooth face of a grain of sand.

Up above, more raindrops were dragging more limestone down to the dune. Speck by speck, a thin layer of limestone formed under the surface of the dune. It was as though someone had built a dune, coated it with a thin plaster skin, and then piled another three feet of sand over that. But it took hundreds of years to plaster over the holes between sand grains.

"Or even *thousands* of years," Loope says. "That's a wild-eyed guess, based on dust-delivery rates. If you had an abundant source of calcium, you could do it faster."

As the sheet of limestone hardened inside the dune, other types of dust were slowly transforming the overlying sand. Each time a raindrop threw down an insoluble speck of dust, it was left to dry on the sand grains. Gradually, these microscopic dusts added up to a coating of clay wrapped around each grain of sand. Slowly, too slowly for an *Oviraptor* to notice, this dune developed a slippery limestone sheet. And over the sheet it developed a heavy blanket of minute marbles—sand grains wrapped in smooth clay coats.

Had the dune been vigorously marching across the desert, none of this would have happened. Collisions among the sand grains would have broken up the limestone layer and knocked the clay coats off the sand. The strong winds that push sand dunes would have blown away the dust. But this was a quiet time in the Gobi. It seemed like an ideal time to raise a family. In the valleys between dunes, ephemeral ponds shimmered. Plants had moved into the dunes, and the minerals in the dust fed their growth. Their small branches slowed the wind and tamed the restless sand. The plants attracted a food web of animals. Dinosaurs and turtles, lizards, mini-mammals, and birds all set-tled into the neighborhood.

Then one day in ancient Ukhaa Tolgod, proposes Loope, it really rains. Never mind about the dust swept down in these drops. The water alone will do the dirty work.

The last time it had rained this torrentially, the limestone layer under the dune's steep face was more porous. The water had squeezed through the holes and continued down through the dune. But since then the slowly accruing limestone has plugged more gaps. The rain rushes down, hits the limestone sheet, and stops. Trapped between the sand and the limestone, water begins to accumulate. Trickles blaze little trails down the face of the buried limestone sheet, but the water doesn't drain fast enough.

Gravity has long urged the face of this dune to descend. But like a child stuck on a dry playground slide, the sand has resisted. Today water lubricates the slide. The face of the eight-hundred-foot dune slips suddenly downhill. A million million clay-coated marbles roll together. Like a snow avalanche, the neat layer of sand breaks up and tumbles toward the base of the dune. As mud-slides can, this wet-sand slide may travel faster than a man can run. As the rushing slurry of sand fans out over the desert floor, it may pick up pebbles, rocks, plants, and animals and tumble them, too.

Its energy waning, the sandslide flows toward Big Mama on her nest. If the sand rumbles or hisses, she doesn't take the hint. One moment she is shaking rain off her back and shifting to cover her eggs. And the next moment a heavy wall of sand is rolling overhead. By now the sand's energy is insufficient to plow the dinosaur off her nest. But its weight will suffice to pin her down—for millions of years.

In the months following the sandslide, bugs may have burrowed down to feed on the carcasses of the trapped animals. And then the fine grains of clay-wrapped sand would have gently filled the gaps between the buried bones, pre-serving even the microscopic ear bones of mouse-size mammals.

Fossilization was a long afterthought: rain continued to ferry melted min-erals down through the sand; and as the buried bones dissolved, each molecule was faithfully replaced with a molecule of rock. Pale dust was transformed, over quiet eons, into white bone shapes. Iron in the clay coats turned rust-red, and this, too, infiltrated some of the bones, to make maroon bone shapes.

If the bones took a few thousand years to fossilize, then up on the surface a change in the weather might be rolling across the landscape. The climate might be cooling. The dunes might be shaking loose their shackles of plants. The animals might drift off to greener pastures. Ukhaa Tolgod might become a ghost town for a while. And then the rain would come again, and the plants would return, and the animals would be lured back in . . . There may be a million years' worth of serially sandbagged animals stacked up here, Loope speculates.

Standing among the bones in the stinging wind, he shuffles a boot. A puff of dust lurches up into the wind. The thin yellow cloud sails north, clearing the top of the valley wall, soaring above the tent village and onward.

5

A STEADY UPWARD RAIN OF DUST

In the first chapter we took a quick census of the world's up-raining dusts. The oceans donate huge amounts of salt, which leaps off the ocean surface in droplets and forms little crystals in the air. Dry lake beds lose great quantities of glass-shelled diatoms. Forest fires produce plumes of black soot particles. Bits of life, from bacteria to viruses, molds, pollen, and fragments of insect, all flow up into the sky. Penguins, and even trees, can send a cloud of chemical beads into the air. Alongside these natural dusts the dusts of human industry rise thickly, and sometimes with deadly consequences. The world over, a constant mist of tiny particles rains upward. But among natural sources few dusts are as dramatic and prolific as the specks that burst out of a volcano.

When the jungle-green hills of the Caribbean island of Montserrat began to rumble in earnest, in 1997, volcanologists around the world began packing their bags. Never mind that the rock stars and captains of industry would soon abandon their vacation homes in Montserrat's famously beautiful hills. Never mind that Montserratian farmers would soon die in their verdant fields. Volcanologists wanted to see the dust fly.

On June 25, Soufrière Hills volcano, which had long loomed green and silent over the tiny farming towns and the pastel capital town of Plymouth, erupted. After a long, slow journey through the Earth's crust, a glob of melted rock finally forced open a crack in the mountain. Gases that had separated from the molten rock were suddenly free to expand when they tore out through the crack. Exploding gas bubbles blasted the mush of hot lava into tiny flecks of ash and foamy balls of pumice that rocketed into the air. When the air could hold no more, the ash fell like a boiling rain on the volcano's flanks. Still squirting thousand-degree gas, the ash rushed downhill at a hundred miles an hour.

Working in their emerald fields between the volcano and the blue ocean, farmers would scarcely have had time to think. The boiling dust buried them, as it buried their peaceful villages, burning everything it touched. When the rumbling stopped, only steeples and stone sugar-mill towers protruded from a gray blanket. Atop the mountain, a small gray-white plume of ash and gas rose innocently into the blue sky.

By the following spring the Montserrat Volcano Observatory was well established, in a once gracious stucco house at the north end of the island. Scientists and students rotated in, stayed a month or more to gather data, and then returned whence they came.

Hayley Duffell was a typical student, from Yorkshire, England. She was slim and wore her thick, dark hair in a sensible ponytail. With a quirky grin she introduced herself as Dust Girl. On one typical day she turned up at the observatory and checked with her colleagues in the seismograph room for evidence of an impending explosion on the mountain. Finding none, she climbed into an orange canvas "hot suit," then got into a waiting truck. David Pyle, the senior-scientist-of-the-month at the observatory and a professor of volcanology at the University of Cambridge, took the wheel, and they headed out to empty the "ashtrays"—strategically placed rain gauges for falling dust.

"Well, some of them are table drawers," Duffell snorts, rolling up her window against the fine gray powder that rises as the truck heads south toward the Exclusion Zone, home to empty mansions, hotels, and farmhouses. She and Pyle pull painters' masks over their faces and wave to a guard as they pass through the barricade. Pyle drops the truck into low and churns up the ash-powdered ruts of a dirt road. Despite the closed windows, chalky dust filters into the cab of the truck. In the valley below, two dead cows are disintegrating on the ash-strewn golf course.

"I'll have to take a four-wheeling course when I get home," Pyle apologizes as the truck bucks in the ruts. Bright-eyed and handsome, he is very polite, very English.

Duffell glances at him. "Take your thumbs out of the wheel. If the wheel spins, you'll lose 'em. That's all you need to know. Intensive course."

Duffell's first ashtray is a small dresser drawer, lying near a hotel.

"They don't mind," Duffell says, stealing a look at the abandoned property, whose gritty guesthouses gaze out over the blue sea. "We'll give it back when it's all over." The air smells of sulfur, and the volcano looms huge over Dust Girl's shoulder. The entire mountain is now the color of cement, although some blackened tree trunks remain like patches of unshaven whiskers. A white plume of dust and gas rolls steadily out of the volcano's ragged throat, floats over the ash-swamped wreckage of Plymouth and out to sea. Pyle holds the drawer and, with a paintbrush, gently herds the fine dust into one corner.

"Mind the hole," Duffell warns, holding a plastic bag to catch the trickle of dust. This ashtray proves to be one of the day's few unmolested ones. The next, a paint can, has been tossed onto a pile of brush on a newly neatened lawn. "Bloody vandals!" Duffell grumbles, replanting the stake that held the can upright. The next tray has been kicked over by a brown cow. Dust Girl shakes a menacing finger at the wide-eyed vagrant. And the next, a plastic bucket, has been upended on a hilltop, where tourists gathered the week before to watch an eclipse of the Sun. "Someone's been *sitting* on it," Duffell yelps. "The eclipsers!"

The last stop is at the observatory's first home, another tropical palace now considered dangerously close to the volcano. Duffell's footsteps on the lawn raise little puffs of ash. The cool tiles of the floor shriek with grit. The pool is greening around the edges and is paved with a dark layer of ash. On the patio Duffell reads a number off a Dust Trak, a small metal box that sits on a lawn chair day and night, breathing through a tube and counting the number of microscopic particles that pass through its electronic lung. The readings have been dropping as the volcano rests.

In dustier times Duffell has clipped miniature versions of the Dust Trak onto human beings for a day, to see how much ash they were inhaling. Even now gardeners mowing lawns on the island are often lost in a cloud of the pale dust.

The ash, geologically speaking, resembles powdered glass, and it is unpleasant in myriad ways. It etches a permanent white fog onto eyeglasses. It harasses the inside of the nose until it bleeds. It stiffens hair into a dull and desperate snarl. "On Montserrat," one geology student was heard to mutter, "every day is a bad hair day." The dust enters houses to finely coat magazines and plates and the delicate innards of computers. Add water to the dust—dew will do— and the dust's sulfur content turns acid and eats the paint off houses and cars. Those few Montserratians who still live in their homes hose them down every morning, to banish the previous night's sprinkling of ash.

But the deepest worry is how the dust interacts with lung tissue. Cristobalite is a crystalline mineral that forms in the baking rocks of a volcano. And when that hot rock is shattered by the next eruption, those crystals shatter into tiny, breathable pieces. Breathe enough and you'll risk silicosis, the crippling disease of miners. So Duffell sends samples of Montserrat dust home to England for analysis. And the word from England has been—and remains— that there wasn't enough cristobalite in the dust, nor dust in the air, to produce an epidemic of lung disease on Montserrat.

But that is just one glassy fragment of the volcanic-dust story. Soufrière Hills volcano is a minor volcano. As the eruptions wound down in 1998, scientists estimated that the volcano had tossed only enough dust to fill 150,000 boxcars, which would stretch merely from New York to Denver. Such a low-powered volcano can't shoot its dust very high into the air. Montserratian dust probably didn't travel much beyond the neighboring islands. A bigger volcano, however, can throw dust a lot higher and considerably farther.

Rick Hoblitt, a volcanologist with the United States Geological Survey (USGS), rotated into the senior-scientist slot on Montserrat when David Pyle rotated out. Hoblitt has the classic look of a geologist, with an impressive mustache and leather hiking boots a team of surgeons will have to remove when he dies. A world-roaming volcano expert, he was deployed to the eruptions of both Mount Saint Helens and Pinatubo.

"In small eruptions the ash only goes twenty or thirty kilometers [twelve to twenty miles] downwind before it settles out," Hoblitt says. "Even in bigger eruptions only a very small amount stays up longer." But that small amount can stay aloft for days—or even years. Roughly once a year a volcano somewhere will blow its stack with enough force to push gas and ash all the way up through the troposphere and into the stratosphere. The troposphere, the lowest seven miles (plus or minus) of the atmosphere, is a busy place, full of wetness and weather. But the layer above it, the stratosphere, is quiet and dry. Dust can circulate there for days, weeks, or years before slowly tumbling back into the chaotic troposphere.

Mount Saint Helens, in Rick Hoblitt's backyard in Washington, certainly made the grade when it erupted in 1980. That catastrophic explosion set loose fifty times more dust than did Montserrat's Soufrière Hills. Most of the dust settled back to Earth quickly, dirtying ten states. But in just fifteen minutes an immense column of steam and gas roared up through fifteen miles of atmosphere, carrying dust with it. Riding the strong, dry winds of the stratosphere, dust from Mount Saint Helens crossed the United States in three days. It circled the world in two weeks.

And even Mount Saint Helens was a decorous little duster compared to the Philippines's Mount Pinatubo. In June 1991 that monster shot out a *cubic mile* of dust. Pinatubo's plume rose twenty-two miles, poking through the ozone layer that lies like sandwich filling in the middle of the stratosphere. Satellites tracked Pinatubo's dust around the planet through several full circuits.

But even a modest blast of dust, once airborne, is sufficient to wreak havoc on an airplane. At the Montserrat Volcano Observatory in 1998, directly above the desk in the seismograph room hung a sheet of paper with the phone number of the regional air traffic control office. It was among the first numbers the volcanologists dialed whenever the volcano coughed.

Just as the little cloud that chugged out of Soufrière Hills's crater appeared snow-white, so do enormous clouds of volcanic ash drifting through the sky. Certainly the crew of KLM Flight 867 gave not a second thought to the white clouds they encountered as they crossed the jagged and snowy Talkeetna Mountains of Alaska in 1989. They had no reason to suspect that Redoubt Volcano, a hundred and fifty miles away, was the source of the thin layer of clouds they planned to drop through on their approach to Anchorage International Airport.

Then they were inside the cloud, and the cabin grew suspiciously dark. Sparks like fireflies whipped past the windshield. As dust and an odor of rotten eggs filtered into the airplane, the first officer attempted a steep climb back to clean air. It was too late. Ramming more air through the engines rammed more ash through, too. And the fact that this ash had been liquid when it erupted just ten hours before was a warning that everything on Earth has a melting point. As the dust was drawn into the hot engines, it melted again, and stuck. It built up in the engines until, one after another, they flamed out. Many of the flight instruments went dead, too.

From 25,000 feet, the plane glided down toward the mountains. The pilot tried again and again to start the engines. In the dark, sulfurous cabin, the passengers were quiet. Just a few thousand feet over the mountain peaks, two engines revived, and the plane stabilized. But to complicate matters, the windshields had taken a severe sandblasting, which added unnecessary excitement to the Anchorage landing. Mechanics later found Redoubt dust in the engine oil, in the hydraulic fluid, in the water supply, in the carpet and cushions, and plugging many instruments. It took two months, four engines, and $80 million to restore the airplane.

The story of Flight 867 is not unusual. Two days later that same ash cloud had wandered three thousand miles south to Texas, where it plugged one engine on a 727, and sandblasted a Navy DC-9. The world around, an average of five jets a year get a dangerous dusting. The aviation hazard of volcanic dust clouds can be amplified by a few factors. Darkness is one. While an ash cloud in daylight looks perilously similar to a normal cloud, an ash cloud at night is simply invisible. Second, networks dedicated to issuing ash warnings are incomplete. Third, even if warning systems were in place to alert pilots, it is not uncommon for a remote volcano to fire off without anyone's noticing until the dust drifts into civilization. And finally, ash can be hard to follow. Once it's riding the wind, it can travel fast and change direction in a matter of five or ten minutes. As a result, even a famously huge volcanic eruption can catch pilots unaware. Pinatubo's plentiful dust, which was spread deep and wide by an ill-timed typhoon, fouled twenty airplanes in three days, inflicting $100 million in damage.

A second kind of volcanic dust leaves airplanes untouched but causes the entire Earth to shiver. When the planet cooled measurably after Pinatubo's eruption, the glassy flecks of ash were not to blame. It was the tiny beads of sulfurous dust that surrounded the Earth and bounced incoming sunlight back out to space.

The Philippine islands, like Montserrat, sit atop a seam in the planet. In fact, the entire Pacific Ocean is underlain by a particularly combative section of the Earth's crust. Its rock-melting conflicts with neighboring plates have earned the region the nickname the "Ring of Fire."

In June 1991 a huge bubble of molten rock found its way to the surface in the Philippines. As sulfur gas released from the magma escaped through cracks in Mount Pinatubo, the telltale stench of rotten eggs rode the air. When the mountain finally gave way on June 15, a gray pillar of hot ash roared upward, accompanied by some 20 million tons of sulfur dioxide. Sulfur-wise, it was the biggest hit of the century. And about half of that gas would soon gel into little balls.

As it ripped up through the troposphere, the gas cooled. Then, in the frosty stratosphere, the chilled molecules began to condense. Over the course of the summer of 1991 the sulfur gas slowly gathered itself into tiny spheres. When there was moisture to be had in the stratosphere, these balls gathered it and turned liquid. When there wasn't, they dried out. But there is never enough water vapor in the stratosphere to form rain, so these balls lingered, riding the high winds. By early 1992 a thin haze of sulfur balls swirled high over most of the Earth.

The balls soon grew large enough to reflect substantial sunlight, bouncing nearly 5 percent more warming rays back into space than normal. And the Earth shivered. The winter was cold, especially in the Middle East. The summer was cool, especially in North America. At the depth of the chilling effect, the world's thermometers, averaged planetwide, registered a degree and one-third below normal.

The little sulfur balls had more tricks up their sleeve. Even as they cooled the Earth, they heated the atmosphere, which changed the planet's wind patterns. They sped up the complicated chemistry that splits apart protective molecules of ozone in the stratosphere, allowing more destructive ultraviolet radiation to reach the Earth. Perhaps the beauty of the sunsets made up for it. Different wavelengths of visible light reflected differently from the volcano's dust layer and drew streaks of purple and red across the evening sky.

But as the balls continued to clump together, they eventually grew too heavy to loiter in the stratosphere. Two or three years after Pinatubo's eruption the sulfur balls began drifting down to the troposphere, where they joined the crowd of other dusts and soon rained back to Earth. Global temperatures normalized in 1993, although some of Pinatubo's sulfur balls lingered high in the stratosphere for another four years.

Soon a fresh cloud of sulfur dioxide would replace Pinatubo's ash and sulfur. In an average year, volcanoes leak or blast out nearly 10 million tons of sulfur dioxide. Pinatubo was clearly a sulfur overachiever, hurling up twice that tonnage in a single day. Montserrat's volcano was more characteristic, contributing just 1 million tons over the many months of its spasm.

For all its airplane-choking and Earth-chilling, volcanic dust is important to the smooth functioning of the Earth's systems. Perhaps if the Earth had never had a sulfurous sky, life would have evolved to use plain, pure rainwater. But, strange as it seems, living things are designed to take advantage of the gentle acidity that soaring sulfur lends to falling rain.

While volcanoes make a big production about throwing their dusts into the air, the oceans are subtler. Even the odor of ocean dust, the salty air, is unobtrusive. But don't let that fool you. Although the fact presents a challenge to the imagination, the world's oceans may rain 3 billion tons of dust up into the air—more even than the deserts do.

The speeding wind rumples the surface of the oceans into whitecaps. Whitecaps are masses of bubbles. When those bubbles burst, little droplets of salt water fly into the air. The water evaporates, and a little crystal of salt remains, airborne.

The air over the southern oceans is particularly well salted, says Patricia Quinn, a researcher with National Oceanic and Atmospheric Administration in Seattle. "There's almost no landmass down there," she says. "The wind really gets going, and there's no land to break it up."

Because, as dusts go, flying salt crystals are often rather large not every speck will travel far before falling back into the ocean. The air of seaside towns is often heavily seasoned. But the air grows less salty over the middles of continents. Los Angeles's air, for instance, is salty. But only the smallest crystals fly all the way to Chicago or Nashville. "The small particles can go for days," Quinn says. "Some of this stuff can get high in the troposphere and travel a long way." Far overhead, at this moment, there may be a pinch of sea salt.

And if that sea salt has a faintly sulfurous taste? Then it's been contaminated with another type of ocean dust. Until 1972 scientists weren't sure where all the sulfur balls in the sky came from. They knew that volcanoes occasionally burp out great quantities of sulfur gas and that swamps and bogs add more. But these stinkers didn't produce nearly enough to account for all the sulfur that washes out of the sky in raindrops.

Then atmospheric chemist James Lovelock went roaming the seas in search of plankton. And in the aftermath of a plankton bloom he found the water awash in sulfur. Phytoplankton ("plant wanderers," in Greek), are single-celled organisms that harness the Sun's energy to grow and multiply. They come in a huge range of shapes, from chains of blue-green pearls to solitary cells plated in lacy armor to the microscopic glass ornaments called diatoms. Estimates of the number of diatom varieties *alone* run as high as a million. And for reasons that remain unclear, some species of phytoplankton carry around a lot of sulfur, in the form of a chemical called DMSP, short for dimethylsulfoniopropionate.

Phytoplankton are always present in the ocean, but they truly bloom when ocean currents deliver deep, nutrient-rich water to the surface. As the plankton population explodes, the water turns milky green, or brown, or even red with life.

And almost immediately a second explosion follows: Zooplankton ("animal wanderers") attack the phytoplankton and multiply. As these marauders devour the phytoplankton, they also seem to alter the DMSP to dimethyl sulfide, commonly known as DMS, which floods into the ocean as the carnage spreads. A day or two after the phytoplankton bloom peaks, the concentration of DMS in the water reaches its own peak. Because phytoplankton depend on sunlight, all this drama takes place in the top few yards of the ocean. And that is where the DMS is left when the bloom goes bust.

Now, when bubbles break on the windy surface, they will release both salt *and* DMS gas to the atmosphere. And high in the air the DMS mixes with other gases, and some of it condenses into particles. Scientists now know that plankton sponsor a wealth of sulfur particles in the atmosphere—perhaps half a million tons of specks a year.

But it's not easy to catch such humble processes red-handed. Sometimes discovering a new source of particles in the air is a matter of luck. University of Hawaii oceanographer Barry Huebert, a soft-spoken man with a warm smile, recalls a fortuitous encounter with "penguin particles." He was on an airplane loaded with scientific sensors, sampling the purportedly clean ocean air between Tasmania and Antarctica. As the plane passed over lonesome Macquarie Island, Huebert's headset crackled to life. From the jungle of instruments in the back, a colleague asked, "Hey, did we pass an urban area? We just saw a huge spike of ammonia!"

Simultaneously, a particle-counting instrument showed a flurry of activity, registering a cloud of extremely tiny particles—really more like big clots of atoms. As the plane flew on, the particles whizzing through the counter grew steadily larger. The ammonia gas was condensing into balls. And the source? Clouds of ammonia usually rise from human and animal breath, manure, and old food. But human beings are hard to come by on Macquarie Island, a thousand miles north of Antarctica and another thousand south of Tasmania.

"Penguins," recalls Huebert, with a little chuckle. "We didn't know then that the stench of ammonia in penguin rookeries literally chokes the people who work in them. Nor were we aware that the zoologists wear high rubber boots when they have to walk around in all that excrement."

By accidentally flying through penguin-poo dust in the making, Huebert and friends had caught the magical transition of gas to particle. It does seem that everything under the Sun is circulating over your head. Plankton fluids and sea salt and penguin poo—that would seem to be more than sufficient in the way of airborne oddities. But every corner of the Earth has a contribution to make. And some of those contributions are alive.

Estelle Levetin, an "aerobiologist" at the University of Tulsa, studies an upward rain of dusts that are neither mineral nor chemical. They are vegetable, bits of life that exploit the air as a means of moving from one point to another on the surface of the planet. "There are two big categories," she says. "Mold and pollen." Each of these contains hundreds of thousands, perhaps a million, individual varieties. "And then there are bacteria. Viruses. I've caught algae and diatoms. And insect parts—wings, hairs, sometimes a whole leg," she says brightly. "Insect parts are very common."

Molds, also known as fungi, cover a lot of the planet's damp surfaces. Consider leaves. Trees excrete all manner of substances onto their leaves, and fungi obligingly clean them up, weaving a microscopic net of rootlike hyphae over the surface. When the fungi are ready to reproduce, they bake up a batch of spores and release them to the wind. If mycologists, or fungus scholars, should one day identify a million different species of fungi in the world, Levetin predicts that 950,000 of them will prove to use the wind to spread their spores. "Some do it passively," she says. "As the wind sweeps across leaves or soil, the spores are just swept off the mold and into the air." Others take no chances. These tiny organisms actually shoot their spores into the air. "Often this requires moisture, so it tends to be a morning activity," Levetin says. "The reproductive structure absorbs some moisture. Then it swells, which creates pressure. And that pressure just shoots these spores out. And you know, people say the air after a rain is really clean? But it's *filled* with spores. After a rain, you have *thousands* and *thousands* of spores."

Because people who sneeze in the presence of pollen have gained a certain amount of clout, lots of agencies monitor the levels of pollen in the air. But the tests to determine if a person is allergic to fungal spores are more difficult to manufacture, Levetin says. As a consequence, that particular clan of "bioaerosols" has been grievously ignored, despite two compelling facts. We sometimes breathe air that contains more than 5,000 fungal spores per cubic foot. And many of these spores are minuscule—smaller than pollen, and small enough to slip into your lungs.

"It's like breathing soup," Levetin says. "No one ever talked to me about how abundant the bioaerosol load is," she marvels. "Now it's a big emphasis for me when I'm teaching. Students have to turn in a specimen collection of mushrooms, puffballs, and bracket fungi. But now they also have to get five mold cultures. And they just walk around campus waving a plate in the air. Then they culture it. It's easy."

Pollen, however, remains the golden symbol of aerobiology. Those plants that rely on insects and birds to distribute their pollen often make big spiny or pitted grains. But plants that exploit the wind make lighter, smoother grains. The stamens of a flower avail the mature pollen to the wind, and the wind whisks it away.

Take ragweed (please!). Lanky and unlovely, ragweed can grow in the roughest and most impoverished soils—the sorts of soils that human settlements often produce. In the United States, August brings the start of the ragweed bloom, although only a loving botanist would appreciate the tiny, weed-green flowers. Each morning these blossoms release a fresh round of pollen. One ratty-looking plant can churn out a billion minute grains. And since this pollen is exceptionally small and light, it can fly for hundreds of miles on the wind. In Tulsa, Levetin routinely traps thick clouds of mountain-cedar pollen, a famously sneeze-worthy species, that have traveled four hundred miles to reach the lungs of city folk. The fact that NASA's cosmic-dust-collecting planes bring back pollen from the stratosphere is an indication of how mobile these particular dusts can be.

And the upward rain of pollen may be thicker these days. The way people use land may have pumped up the pollen volume, especially in the West. "There has probably been an increase in pollen in the West since about a hundred years ago," says USDA range-land ecologist Dennis Thompson. That's when ranchers grossly overstocked the western range to feed a booming human population. The cows ate everything in sight—or almost everything. "If you allow it, the animals will overgraze the plants they like the most," Thompson says. "And another plant, that's less preferred, will take its place." The plants that usually invade an overgrazed field are annual plants, instead of perennials whose roots survive the winter.

"And if it's annual, it needs a lot of seeds," Thompson says. "So it needs to make a lot of pollen. Goldenrod. Ragweed." Thompson proposes that shifting cattle from pasture to pasture, timed to the growth stages of weeds, could injure the pollen makers and reduce their productivity. Ranchers have yet to test this method of dust control.

Even if they did, global climate change may prove to be an even more powerful producer of excess pollen. When the U.S. Department of Agriculture recently experimented with growing ragweed in a greenhouse spiked with extra carbon dioxide (the dominant global-warming gas), they found that the plant produced more pollen. Researchers at the agency suspect that the world's ragweeds have *doubled* their pollen output in the past century and think they may do so again in the coming century, as extra carbon dioxide continues to alter the atmosphere.

What sort of tonnage do fungal spores and pollen add to the air? No one knows. But another strange plant dust is known to contribute hundreds of millions of tons each year. The chemical compounds in these particles led some members of President Reagan's administration to direct their environmental passions against shrubbery. And indeed, if the tiny beads of chemicals that rise from forests and lawns were to emerge instead from a factory, many would be called pollutants. But scientists still aren't sure why Reagan's "killer trees" give off chemicals like isoprene, terpene, alcohol, and formaldehyde.

"Isoprene is released from deciduous trees, mainly," says Brian Lamb, a chemist at Washington State University. "It is only released in the presence of sunlight, in the daytime. So maybe it's a response to temperature stress. Terpenes tend to come from conifers, and that may be temperature-driven as well. Hexanol—you know that newly cut lawn smell—that may be an injury response."

Whatever the reason for these chemicals, the effect is the same: plants the world over release a veritable fog of these chemicals into the air. Only a fraction of this vapor actually condenses into tiny blobs or sticks to other grains of dust in the air. The distinctive vapors of pine, citrus, and mint are among those likeliest to form solid particles in the air. And they add up. The Great Smoky Mountains are not smoky due to fires. Some of the famous haze is indeed beads of pollution. But much of the "smokiness" is the work of natural tree beads. If the Sahara exports about a boxcar of dust every four seconds, all the world's plants may exhale a boxcar of particles every eight to twenty-four seconds.

Killer trees? It is probably more realistic to view tree dust the way we view volcano dust and other natural particles. They are inevitable. And what's more, the planetary community has probably developed some creative uses for these complex plant specks. The question that interests Lamb and his colleagues is how the world is adjusting to the fact that such human endeavors as dry cleaning, shipbuilding, and electroplating add huge quantities of similar chemicals to the air.

Some bioaerosols are entire, living (or dead) things. Flying diatoms don't add significantly to the airborne vegetable matter, in terms of simple tonnage. But when these glass-shelled algae do take a spin through the atmosphere, they raise interesting questions. They seem to defy the size limit for far-flying dust, for one thing. And they may sometimes fly with a purpose.

Michael Ram, a professor of physics at the University of Buffalo, has become an expert at teasing these tiny organisms out of ice cores from Antarctica and Greenland. Deep glaciers preserve thousands upon thousands of fine layers, each representing a year. And trapped in each layer is a sprinkling of fallen desert dust, stardust, volcanic ash, pollen, insect parts—and diatoms. Ram melts a bit of ice, then puts the remaining sediment under a microscope. The diatoms, he says, stand out due to their geometric perfection. Desert dust, under the microscope, resembles shattered rock. But diatoms often resemble delicately etched pillboxes or broken shards of the same.

Most diatoms spend their brief lives adrift in rivers, ponds, lakes, and oceans. And when they die, their little shells sink. Ram says the ideal source of diatom dust is a shallow lake that shrinks in the dry season, exposing the sediment at its edges to the wind. Africa and the western United States are both pocked with excellent candidates.

Ram originally intended to use the diatoms he found to trace the source of the dust and diatoms in each sample: if the ice of one century was rich in North American diatoms, and the next century's ice held African diatoms, he could conclude that the prevailing wind had shifted. This might reveal something about the dynamics of climate change. But Ram's diatoms proved coy about their place of origin. Many of them look alike. Scientists with more diatom expertise are pursuing this line of inquiry.

And Ram's diatoms have caused additional head scratching. Generally, scientists don't expect things much larger than a few hundredths of a hair's width to fly long distances. But Ram has seen disks as wide as a hundred, or even two hundred, microns—that's a whopping two hairs in diameter. "These diatoms are large, but they have a large surface area, and they're light," Ram speculates in the accent that remains from his Egyptian upbringing. "They're like Frisbees. They're very aerodynamic."

The size of the diatoms may also relate to the strength of the wind that lifted them. An uncommonly strong wind can lift uncommonly large dust, as a survey of hailstone cores has suggested. Carried up into a storm cloud and then coated in ice until they fell again have been such "dusts" as small insects, birds, and at least one gopher tortoise. Perhaps a large diatom is not such a challenge.

But a third source of puzzlement is what appears to be a complete colony of diatoms that evidently dwelled smack atop the Greenland glacier about four hundred years ago. It is common for living diatoms to blow into melt pools at the *edges* of glaciers and there start a family. But the founder of the little clan Ram discovered apparently flew all the way to the center of the immense island before dropping into a puddle. And that pioneer was still in good enough shape to launch a modest dynasty.

"Usually, it's clear that the diatoms were windblown *after* they were dead," says Ram. "They're mostly broken, fragmented. But when we looked at these diatoms, we saw like a family—same species, same size, everything. The size tells you they all thrived for the same amount of time. So here we must have had some diatom that was still ticking, okay?"

Okay, a pinpoint's worth of identical diatoms found high on a glacier is, scientifically, proof of nothing. But it does suggest that some diatoms have evolved to make the most of their "blowability."

Diatoms would not be the first whole, living organisms to hop around on the wind, of course. Nor would they even be the largest. The harshest and driest deserts on Earth, Antarctica's McMurdo Dry Valleys, are not lifeless. Granted, the "top predators" are nematodes that stalk and kill bacteria. On warm days these regal—if barely visible—worms survey their domain from inside the water film that coats soil particles. And how did these kingly worms travel to the remote Dry Valleys? They seem to have come on the wind. One scholar has suggested that since glaciers essentially scoured Antarctica clean during the last ice age, *many* of that continent's small beasties must have blown in from other continents after the ice retreated. The size limits on flying life remain unexplored.

The flight restrictions that nature places on viruses and bacteria are better understood. Fortunately, the virulent and aggressive germs are generally fried before they can carry their diseases very far. "They get knocked out very easily," says Milton Leitenberg, an arms-control specialist with the Center for International and Security Studies at the University of Maryland. "It's not just UV and oxygen. They also dry out. In some cases it's a matter of minutes, or even seconds. Now, anthrax is an exception."

Anthrax is a common soil bacterium whose spores travel in tough, protective coats. In its natural cycle, anthrax mainly threatens people who work in direct contact with infected animals. In its *unnatural* deployment as a weapon, a package of bacterial spores would be hurled at the enemy. But still, only those people in the vicinity would inhale it. In 1979 a Russian anthrax plant sprang a leak, freeing a plume of spores. Near the plant, ninety-six people became infected, two-thirds of whom died. Thirty miles downwind, as the plume became diluted, only sheep and cows sickened. And even this sturdy germ, if it spends too many weeks or months in sunlight, becomes just another bit of lifeless flotsam in the wind.

More exceptional may be a race of bacteria that do indeed live in the air. Austrian researchers have recently discovered that the cloud droplets they capture in the Alps are rife with living bacteria that appear to be reproducing on the fly. Noting that the cloud droplets the bacteria inhabit would shield them from UV, the scientists soberly concluded that much of the Earth is covered by clouds, and these clouds should be considered as "habitat."

Increasingly, health experts are interrogating the tiny inhabitants of the air, both the living and inanimate, about their intentions in the human lung. As we'll see in a later chapter, asthma rates are soaring worldwide, and the search for an airborne cause is becoming intense.

Right now, Estelle Levetin says, many scientists don't even know the term "aerobiology." But this is a growth industry. Levetin will not be lonesome for long.

If some young scientist should hanker to carve a name for herself in a field even less crowded than aerobiology, she should know that aerobiologists ignore some of the world's most interesting dust*making* organisms.

Lichens, for instance, those crinkled cooperations of algae and fungi, make a living by slowly pulling nutrients out of the rocks they inhabit. Their fibrous hyphae invade fractures in the rock, where they excrete acid to dissolve out minerals. As the rock weakens, dust flakes off. In addition to chemical weapons, lichens also use their sheer strength to break up rocks. Jammed in a crack, their hyphae shrink in dry times and swell in wet ones, forcing open the fractures in the same manner as do salt and ice crystals.

Even less appreciated are the microscopic bacteria and fungi that "eat" rocks. Together, lichens and microbes act as the sodbusters that venture into virgin rock and, in the course of harvesting minerals for themselves, slowly take the rock apart for others to use.

But perhaps the most remarkable living dustmakers are those probable descendants of the dinosaurs, the birds. The exact tonnage of the world's bird dust is still a doctoral dissertation in search of a student. But, as researchers at the United States Geological Survey (USGS) learned the hard way, it's enough to mess up an experiment.

These scientists thought they had considered everything when they set out an array of "dust traps" in the western United States in the 1980s. They intended to measure how much dust falls from the sky. The Teflon Bundt-cake pans mounted on posts were painted black, to speed the evaporation of rainwater, and they held a layer of marbles, to protect settled dust from the scavenging wind.

But the investigators didn't think of what alluring perches their dust traps made for birds in a landscape devoid of trees. Nor did they consider the "crop stones," pebbles that birds swallow to grind their food. In the course of grinding food, these crop stones themselves are gradually ground to dust and passed through the gut. The resulting "bird-derived sediment" confused the picture in the cake pans: the dust of diminishing crop stones doubled and tripled the apparent dust-accumulation rate.

"And the birds do more than that," says Marith Reheis, a veteran dust trapper with the USGS in Denver. "I believe the ravens actually trade. They take my marbles and leave me things in exchange. Often they leave rocks the same size as the marbles. But a couple of presents I've had were the desiccated head of a lizard and the hind end of a kangaroo rat . . . oh, and a burro plop." Antibird modifications have been made.

If the sky is a naturally dusty place, aswirl with desert dust, volcanic ash, sea salt, tree ooze, fly legs, diatoms, and so on, it is also an *unnaturally* dusty place.

Fire has always been a source of black dust in the sky. Before the primates came down from the trees and began whacking rocks together, the power to set fires was held by lightning—and by rocks that whacked *themselves* together. In South Africa it is still perfectly normal for tumbling rocks to shed sparks and set impressive grass fires. But ever since *we* learned to whack the rocks together, we've been using fire to control the plants and animals that sustain us. We're now the leading cause of fire dust.

Dolores Piperno, an archaeobotanist with the Smithsonian Tropical Research Institute in Panama City, Panama, studies ancient arson. A no-nonsense woman with short blond hair and a firm jaw, she uses the fine layers of mud in lake beds to reconstruct ancient environments. And she cherishes no romantic vision of Panama's Paleo-Indians as environmentalists.

"These were sophisticated hunter-gatherers," she says of the people who populated Central and South America eleven thousand years ago. "They knew how to control fire. They came through here and just started lighting matches."

Her evidence lands with a thud on her lab bench. It's a chilly black cylinder of mud wrapped in plastic and tape. Embedded in the layers is a code-cracking kind of dust called phytoliths. Phytoliths are, literally, "plant stones." Many plants manufacture microscopic rocks in the cells of their leaves, fruit skins, and seed husks. They probably serve as a glassy deterrent to marauding caterpillars and the like. When bran cereal scratches your throat, Piperno says, you're experiencing the fury of phytoliths.

These dusts come in a dazzling variety of shapes, from simplified butterfly and flower forms to dumbbells, golf balls, corn kernels, and even strips that look like ruffle-edged lasagna noodles. Though many fall into the tenth-of-a-hair's-width range, some "lasagna noodles" can stretch to ten times that length—which to a gnawing insect must seem as appetizing as a pane of window glass. And each little rock is packed with information. Each shape is manufactured by a different species of plant. Because phytoliths form around a fleck of organic matter, they can be radiocarbon-dated. And when they burn, they turn black.

So when Piperno isolates phytoliths from ancient lake mud, she can paint a clear picture of how ancient people altered the land. Up until about eleven thousand years ago the phytoliths of forest plants—trees and shrubs—dominated in the mud. And then they began to dwindle. Within about four thousand years many of those forest phytoliths were missing completely: the forests had been eradicated.

And in their place? As the forests retreat, the phytoliths of open-ground plants—sedges and grasses—surge forward. And often these tiny stones are blackened. By four thousand years ago, says Piperno, the grass now known as corn was spilling its phytoliths with regularity. Phytoliths of corn, and squash, too, grew steadily in number—and in size, reflecting their domestication. Judging from these new dusts, Piperno says that agriculture was widespread. And to keep the land clear and release nutrients to the soil, these farmers were setting fire to the land every few years.

Then, after hundreds of years of absence, the forest phytoliths make a sudden return in the mud of five hundred years ago: when the Spanish invaded Central America, Piperno concludes with a rueful smile, the fire farmers themselves were eradicated.

Fires, however, remain central to tropical farming. From the savannas of Africa to the forests of South America, farmers routinely unleash fire to clear the way for agriculture, cow pastures, new roads, and towns.

How much of the torched greenery is carried aloft as gas and how much as soot is in dispute, however, since scientists are just delving into the subject. Grass fires on the African savannas, for instance, are hot and, hence, relatively clean. Cooler fires, in thick, damp forests, generate more smoke and a wider variety of complex chemicals that coat the little soot particles. And burning animal dung, of which some 118 million tons are used for cooking and heating each year in Asia alone, also tends toward murky smoke.

One general estimate of smoke production is that a square mile of burning land will generate a ton and a half of smoke particles every hour. A competing estimate suggests that the rate is more like twenty tons an hour. Either way it's a lot. Smoke plumes are easily seen from space, where they look like broad gray jet trails, streaking across hundreds of miles of land and sea.

And plants outside the tropics are certainly not immune to the human urge to incinerate. In what the fire historian Stephen Pyne has called an "unapologetic addiction," many U.S. farmers persist in the ancient practice of searing the previous year's stubble before planting anew. Health concerns, voiced with increasing alarm, are finally starting to dampen the burning zeal in this country.

The northernmost forests are another a surprising source of soot. Canada and Russia together are home to nearly one-fifth of the world's forests. As in the tropics, many of the northern fires—about half, in Canada—are started by people. And fires in these dry, remote forests often burn unchecked.

In Canada, and perhaps in Russia as well, almost all the fires are "crown fires." These are superhot fires that burn everything from the soil to the crowns of the trees, releasing far more smoke, and shooting it much higher into the air, than a fire that scurries along the ground. In an especially dry year, fire can

even burn down through three feet of the forest floor, which produces an especially thick smoke. Since climate change seems to be heating northern regions
the fastest, there's a chance that these forests will become drier still and will
blow increasing amounts of smoke.

What with savanna fires, agricultural burning, and forest fires, NASA estimates that between one and twelve Texases' worth of land burns every year, destroying between 2 and 11 billion tons of plants and showering many millions
of tons of soot up into the sky. If the heat of a fire lifts smoke high into the troposphere, the soot can travel great distances. Smoke from Mexico, for instance, has been known to tickle the noses of North Dakotans, Wisconsinites,
and Floridians. Fires in Canada often ship rich plumes south over New England or even down to Louisiana.

Despite the heavy human influence, the dust of forest and grass fires is still
composed of the same natural soots and organic compounds. Since there were
first plants to burn, lightning has been launching clouds of this dust.

The fires of war, however, produce strictly human and unnatural soots. Retreating from Kuwait, Saddam Hussein's army reinvigorated the question of
battle ethics when they set fire to 613 oil wells. In addition to freeing great lakes
of oil, the fires sent untold quantities of soot and other chemicals into the sky.
A dark cloud, spitting black rain, spread over Kuwait, then expanded over
neighboring nations. As the fires raged, the temperatures under the soot
plume fell by as much as eighteen degrees.

These are dramatic fires, these raging forest fires and vicious fires of war.
But even the tame fires we set every day put dust into the air. Every match we
strike releases soot. Every hamburger we slap on the grill and every campfire we
circle releases more fine black dust.

Humanity's dusty history began with fire. But now nothing symbolizes our
dusty effect on the world quite like a car. The most obvious kind of car dust results from burning gasoline or diesel fuel. The torching of fossil fuels is rarely
a wholesome activity. They're packed with complex compounds, and they require high temperatures to burn completely. So truck and car engines shoot
plenty of partially burned fuel out the tailpipe.

Diesel engines—typically found under the hoods of long-haul and delivery
trucks, construction and farm machines, buses and trains—are particularly
dusty. City air is often tinted blue with specks of the half-baked fuel, and this
soot settles like black powder on city windowsills. The federal Environmental
Protection Agency (EPA) estimates that one old diesel truck can produce eight

tons of soot and fumes in a year. Even a light-duty diesel engine cranks out thirty to a hundred times more particles than does a regular gasoline engine.

The heart of an exhaust particle, whether it's gasoline or diesel, is a nugget of carbon soot. But diesel soot is coated and contaminated with hundreds of other by-products of combustion. Health researchers are increasingly concerned that these tiny grains cause a range of health problems, from cancer to lung and heart diseases.

Gasoline engines may be less dusty, but they are far more numerous. In the United States alone, they release well over a million tons of hazardous air pollutants every year, ranging from mercury and lead to benzene and arsenic. Although much of this stew emerges in vapor form, it can condense on the fresh-baked soot and other dusts in the air, to build very tiny toxic bombs.

All these engines also release clouds of sulfur and nitrogen gases. And just as some of the sulfur gas from a volcano condenses to form particles, so do the gases from your car engine. In the United States, cars, trucks, and off-road vehicles send up a million tons of sulfur gas to the sky each year—about one-tenth the amount kicked up by all the world's volcanoes. And motor vehicles are the preeminent industrial source of nitrogen gas in the United States.

The puffing tailpipe is an obvious offender. But take a closer look at your automobile. As your tires wear down, where does the rubber go? Tiny crumbs of rubber form an invisible cloud behind your moving car. The EPA concerns itself with counting only the smallest pieces of this dust, the grains narrower than one-tenth of a hair. And of rubber dust that size, cars in the United States put more than 25,000 tons into the air each year. That's equivalent to 2 million tires, powdered.

Friction between the rubber and the road moves your car forward. Friction between the wheels and their brake pads brings you to a stop. Hitting the brakes, in the United States alone, produces more dust than disintegrating tires do: about 35,000 tons of brake-pad dust billow into the air each year. Since the pad materials have diversified from old-fashioned asbestos, brake-pad dust is growing more colorful. It might contain a host of metals, as well as ceramics, carbon, Kevlar, and fiberglass.

Look at trucks and cars from a distance, especially on unpaved roads, and you'll see another enormous source of dust. Vehicles traveling the dry roads of the western United States are often visible first as a billowing cloud of earthen dust. Though dirt roads enjoy a reputation for rural simplicity, they are increasingly vilified by clean-air experts. As a source of dust, they are prodigious. Each year the EPA tallies up the dust tossed upward by dozens of sources in the United States—trains and planes, wild fires and wood stoves, coal mining and cement making, and so on. Together, all those sources put about 33 million tons of dust into the air annually. And of that dusty mountain, dirt roads are responsible for a full third.

In days of yore the popular solution to dusty roads was to spray them with used engine oil. As Exxon *Valdez*–ish as it sounds, it is still practiced in developing nations. But those most familiar with the drawbacks of oiling dusty roads are probably the former citizens of Times Beach, Missouri, a little ex-town twenty miles southwest of St. Louis.

Between 1970 and 1972 a contractor in the road-oiling business added the dioxin-tainted hazardous waste of a pharmaceutical company to his usual brew of used engine oil. This mixture was applied liberally in Times Beach, on unpaved streets and parking lots. Ten years later the EPA discovered that the roads, shoulders, and ditches of Times Beach were steeped in dioxin, a powerful carcinogen. The citizens were evacuated and paid off. Their houses were destroyed, and the dirt roads were shoveled through a hazardous-waste incinerator hauled in for the occasion. In retrospect, perhaps a little road dust wasn't so bad. Missouri promptly outlawed the use of old oil for dust suppression.

Paving campaigns are becoming more common. But paving can run to half a million dollars per mile, which raises resistance in those towns responsible for fixing the roads. Maricopa County, Arizona, for instance, which is home to dusty Phoenix, is responsible for seven hundred miles of unpaved public roads. Even if all those miles were paved tomorrow, dust would continue to billow up from more than three thousand private miles of dirt road in the county. And even paving won't prevent cars from raising a cloud of soil dust. Paved roads put 2.5 million tons of dust into the United States's air each year.

All the soot and the sulfur beads, the brake pads and the tire bits and the plain old soil dust—these dusts are fairly damning of the cars and trucks. And now an atmospheric scientist from Taiwan has shown that even if a car produced no dust of its own, it would still have a tremendous impact on the dust population in the air. Car engines terribly transform the preexisting dust that's drawn into their engines.

Inside a hot engine, says Jen-Ping Chen, the dusts are vaporized. When this vapor leaves the tailpipe, it recondenses as tiny particles that are easy for people to inhale. It's akin to running over a glass bottle with a lawn mower: driving a car breaks relatively safe dust into lots of alarming little pieces.

Chen's work illustrated how these new specks take shape at lightning speed behind an automobile. Measuring the concentration of particles just four inches behind a car's exhaust pipe, he found that one cubic inch of air held about 200,000 particles. And by the time the exhaust was sixteen inches from the pipe, the population of a cubic inch ballooned to more than 6 million tiny reincarnated specks. In damp air, Chen says, these particles will be damp, even liquid. In dry air, auto exhaust is soon as dry as . . . well, dust. Either way these recycled dusts are supersmall, and superbreathable.

Even an electric car produces polluting dusts, indirectly. In fact, any electrical appliance that's wired to the standard electrical grid adds the dust of fossil fuels to the air. When you turn on the coffeemaker, chances are that fossil fuels are roasting somewhere in the distance, to produce the juice. And when fossil fuels are set afire, dust is inevitable.

Coal is the dustiest fuel—and wealthy nations burn a lot of it. A whopping 90 percent of all the human-forged sulfur beads rise over the industrialized nations of the Northern Hemisphere. In the skies of the world these human-launched sulfur beads now outnumber the naturally made beads by two to one. And this horde has transformed acid rain from a natural necessity to a corrosive curse in many parts of the world. Particle-forming nitrogen oxide gases also escape in huge quantities from burning coal. Coal smoke is rich in toxic mercury and other metals, and the radioactive radium and thorium loosed from burning coal have been compared to the waste of a nuclear power plant.

But *everything* we do that involves the conflagration of fossil fuels will throw dust into the air. As an airplane slices the sky, it spills a trail of dust, along with quick-condensing gas that makes more particles. As ships chug across the ocean, they leave trails of dust in the sky that for reasons unknown sometimes linger for days. Rockets, befitting their lofty status, spit sapphire dust, the product of aluminum-fuel combustion, into the atmosphere.

These dusts of human enterprise join nature's own tiny specks for the journey of a lifetime.

6

DUST ON THE WIND HEEDS NO BORDERS

In April 1998, Dan Jaffe, a professor of environmental science at the University of Washington in Bothell, looked up and thought, Something is going on. "It was a beautiful, blue-sky day," he recalls. "Or it was a day you'd *expect* the sky to be blue." A storm had come through the day before, which should have dragged in clean, clear air. "But I looked up in the sky and saw milky, *washed-out* blue. My first thought was of a volcanic eruption."

Not entirely by chance, Jaffe had air-sampling machines running atop Cheeka Peak, where Washington State juts into the Pacific Ocean. He was hoping they would record an ancient story: Asia exporting herself to the Americas. But he had no reason to expect the monstrous size of this particular chapter.

Rivers of dust flow around the world, riding the invisible currents of the air. They are such an integral part of the planet that without them, rain and snow would be rare. But now, as scientists map these subtle rivers, they're troubled by a human addition to the natural dusts. The dust rivers are becoming dangerous. And they flow from one nation to the next without discrimination.

A year after his earthbound machines caught a rich whiff of Asia, Jaffe has outfitted a small airplane with air-sniffing machinery, and this craft is plying the skies over the Pacific, looking for more deliveries of dusty air.

"Where's the cleanest air in the Northern Hemisphere?" Jaffe asks as he drives his muddy white Toyota to an airfield north of Seattle. He waves a hand at the Washington sky. "We think it's probably here. About as clean as we get is a hundred particles per cubic centimeter." However: "We've seen haze—with the instruments—on four days so far," he says, parking and pulling his backpack from among a clutter of children's toys in the backseat. "It definitely comes from Asia."

Already at the hangar are pilot Mark Hoshor and one of Jaffe's gradu-ate students, Bob Kotchenruther. Kotchenruther, ruddy and blond, is busy cranking chunks of chemical ice through a cocktail ice grinder. Inside the air-plane the twelve seats have been replaced with sharp-edged banks of instru-ments, some of which are already humming in preparation for the day's work. One is prebreathing oxygen from a tank, to keep it free of Earthly contam-inants.

Hoshor, dark-haired and reserved, climbs into the rear of the plane and si-dles up the aisle between banks of instruments and computer screens. Crouch-ing under the low ceiling, he points out a safety feature—an inflatable life raft in a plastic bag. Kotchenruther shuffles up the aisle and wedges himself into a sitting position among the instruments. Jaffe, not a fan of flight, waves good-bye as the little plane hums down the runway and into the once-famously clean air of Puget Sound.

Immediately, Kotchenruther's computer screen begins to number the invis-ible millions outside, drawing jagged lines as the particle population rises and falls. In the newest of languages, the computer transcribes one of the Earth's oldest sagas, the story of restless rock.

Springtime dust storms are common in Asia, when strong winds scramble across the dry land. But in April 1998, meteorologists concluded, the winds were especially violent. In the wide deserts and loess regions of Mongolia and north-central China, cold winds may have hit forty-five miles an hour, early on the morning of April 15, 1998. A wind one-third that speed would have sufficed to lift dust. By noon of April 16 the skies over Beijing darkened. Then fat drops of dust-clotted rain spattered cars and streets with desert-yellow mud. This was neither abnormal nor extraordinary. Yellow rain, also called "mud rain," is increasingly common, as overfarmed loess soils turn to bleak deserts. Such massive dust storms used to afflict Asia once every seven or eight years. Now they're an annual event. In the spring of 2000 a dozen smaller events dumped dust onto Beijing.

The yellow rain washed only a fraction of the dust river down from the sky after that powerful storm. Photographed from space by a satellite, the remain-ing dust formed a smooth, coffee-colored river that swirled steadily east, across the blue-and-white face of the planet. It crossed the Pacific and licked daintily at British Columbia before breaking up and fading. And behind it a second storm was brewing.

On April 19 another violent wind dragged its claws across the Gobi. In the rushing dust, visibility fell to as little as fifty yards. The CNN news service reported that the east-ripping storm killed twelve people and nearly nine thousand cows in China and left a thousand Mongolians homeless. By April 21 the tongue of this new river had surged out over the Pacific.

These innumerable grains of dust were entering a wild new chapter in their life story. Some of them had been snatched up from restless sand dunes, inside of which they had been tumbling for millions of years. Others were liberated from the ancient cemetery of the Gobi dinosaurs. Some of the grains were gathered from dry streambeds, having tumbled down from the primordial mountains.

But this river was not pure rock dust. The wind discriminates only according to size. When it roared across deserts, it gathered fragments of everything that can fall apart: bones of camels and horses, crumbs of white fossil, shreds of the colored silk hats of Mongolian herders, papery fragments of dead wild onion, ashes from a campfire of gnarled wood, shreds of camel-hair rope, fibers from a yellow tea carton left to honor a sacred pile of rocks in a mountain pass, and perhaps even a bit of old Genghis Khan himself.

And when it howled past factories and cities, this wind peeled up the corners of China's thick blanket of pollution. Tributaries of sulfur dust and diesel soot, gases and toxic metals, rose and entwined themselves in the swirling river.

Not just China but the entire Asian tiger is turning out to have bad breath. As the economies from Singapore to Thailand, Korea to China, have grown, these countries have developed a ravenous appetite for fossil fuels. Unfortunately, technologies that filter the resulting smoke are expensive. And so the romantic views of Bangkok are often draped in a haze of particles that make the eyes water and the throat clench. In Hong Kong's harbor, ships sometimes rely on their foghorns as they negotiate a sepia cloud of gas and dust that can drop visibility to less than a mile. But in China the fossil-fuel frenzy, combined with an enormous population, produces epic quantities of pollution dusts. And the forecast is very, very dim.

China feeds its tiger largely on coal, and soft coal at that. Soft coal is notoriously dirty to burn, unleashing the customary cocktail of soot, radioactive elements, and poisonous gases, plus a bonus shot of sulfur. Beijing, even without the occasional contribution of deserts, has many times more dust in its air than do U.S. cities. "Beijing fogs" are becoming more frequent, and these pollution mists are sometimes thick enough to cause rashes of traffic accidents.

Coal feeds China's industrial fires, and to make matters worse, small coal stoves heat homes and cook the meals across the country. Thus, even little villages wrap themselves in a coal-dust fog. Air-pollution monitoring often reveals two spikes in the daily dust count: one at breakfast time and a second at the dinner hour. Slowly, China is trying to guide its tiger toward cleaner fuels. But in such a huge nation that will be like trying to turn a supertanker: it won't happen quickly.

In addition to soft coal, soot-rich diesel fuel enjoys tremendous popularity in China. A tradition of burning agricultural fields pumps still more smoke into the sky. And factories whose by-products include toxic dusts like mercury and lead have traditionally spent little money to keep those dusts under control.

Throughout Asia, scientists predict, the pollution cloud will grow at 4 percent per year, which would double the murk in about two decades. In less than half that time some observers predict that Asian winds will start making regular and measurable shipments of ground-level ozone, a key component of smog, to the West Coast of the United States.

But forget for a moment the unsavory products China's winds are *exporting*. The Chinese people themselves breathe these dusts full-strength, before they're diluted in the atmosphere. Noxious dusts in China kill one out of every fourteen people. Pneumonia, often instigated by breathing dirty air, is the leading cause of death for Chinese children. All told, about a million Chinese people die each year from the deadly dust in their air. It is as though the entire population of Maine died of dust poisoning year after year after year.

An unexpected side effect of China's dust blight is that the nation's crops are flagging. A study funded by NASA recently found that the pervasive dust actually casts shade on the plants. On the majority of Chinese agricultural land the harvest may be stunted by 5 percent; under the sootiest stretches the food supply may suffer a 30 percent shrinkage. The discovery has sobering implications for all hazed-over farmlands, including swaths of India and Africa, and the East Coast of the United States.

And in 1998 a tendril of this industrial haze entwined with the desert dust, and rolled toward the Pacific Northwest.

Mark Hoshor steers the dust-sniffing airplane west. He exchanges bursts of syllables with air traffic controllers, as they work out a flight path. A problem had come up before the plane left the ground.

"As I understand it," Jaffe had said, "the military owns most of the airspace off the coast. And sometimes they let us use it. Today the military has said we can't be very near the shoreline." He had grimaced. Then reconsidered.

"Actually, it works out well. There are some easterly winds today, which give us some pollution from North America. Hopefully, as we get farther out, the winds will shift more to the north and northwest." From Asia.

And so Hoshor and Kotchenruther settle in for a long flight out to their block of sanctioned air. As the plane rises above the glaciated Olympic Peninsula, dust is obvious to the naked eye. Off to the northwest a layer of dark bands, as faint as watercolor brushstrokes, traces the curve of the Earth. To the north the bands are blacker; to the west they look more yellow-brown.

"*Natural* aerosols aren't very light-absorbing," Kotchenruther says from his perch at the computer. "Pollutants are blacker. The dark color is suggestive that it could be anthropogenic—or from biomass burning." That's scientese for fires.

As the plane buzzes on, the lines on the computer graph bounce, a blue line recording patches of ozone, a yellow line noting clouds of microscopic particles.

There is always some dust in the sky. A minuscule amount of lead hangs in the air, because there is lead in rocks, and rocks fall apart and blow away. Faint traces of mercury and radon taint the air for the same reason. Cadmium, thallium, and indium leak from volcanoes. Selenium rises from breaking bubbles in the ocean. And an embarrassment of silica and aluminum, calcium and iron fill the air, because those are more plentiful components in rock. Thus, even in the world's cleanest air, any given cubic inch might host 1,500 specks of various pedigree. But west of Seattle the yellow line bounces around well above that baseline. Humanity has spiced this air.

It takes Hoshor an hour to reach his designated block of air. In the distance the watercolor bands still float above the horizon. He guides the plane toward the bottom of the box. As the plane drops through a hole in the cloud layer at a thousand feet, the yellow line on the computer graph leaps, registering a few times more than the normal background level of particles.

"It can't be sea salt at that altitude," Kotchenruther says. "There's cold air trapped at the ocean surface and warm air aloft. There's no mixing." His instruments cannot identify particles on the fly. The exact nature of these specks may emerge from the data later.

A mere thousand feet above the frothy Pacific, Hoshor steadies the plane and points toward Asia. A "leg" begins. Flying straight and level, the plane samples the air for twenty minutes. Kotchenruther flips switches. When the leg is complete, Hoshor guides the plane up two thousand feet, turns back toward Washington, and flies another.

On the fourth leg, more than a mile high in the clear blue sky, the yellow particle line is jumpy right from the start. The wind, as Jaffe predicted, is now out of the northwest. And although the air couldn't look cleaner, the plane is in a cloud of dust. Near the end of the leg the yellow line spikes at about thirty times the clean-air level.

Kotchenruther shrugs noncommittally. Until the data are analyzed and this air parcel's past is reconstructed, there's no way to know whether the cloud consists of Mongolian dust, Alaskan volcano sulfur, or the condensing gases of a plankton bloom. Kotchenruther has been doing this for weeks. And when you're conducting a roundup of the invisible, it's a sad fact that a yellow zigzag on the monitor is a fair representation of the apex of the drama.

The conversation turns to whales, to the teensy cargo ships crawling over the ocean below, to memorable plane rides. The headphones pinch. The bladders balloon. Kotchenruther unwraps a sandwich. Hoshor eats a few cookies. The computer counts and counts.

"I don't know, Mark," Kotchenruther says, studying the computer's rendering of the flight path. "There are little wavers in that last leg. Not your usual protractor-straight line."

"Seventy-knot wind," Hoshor grumbles. "There's gonna be a curve in this leg, too."

In midafternoon the plane finally turns east for the long run back to the continent. Kotchenruther sighs from his perch. "I think I'll take the life raft home," he says, poking the plastic bag. "I want to know if that thing works."

Jaffe will later conclude that the yellow spike was a cloud of humdrum domestic particles, probably swept off North America by a circling wind. But in the five weeks of flights the instruments picked up Asian dust at least once a week.

When the river of dust and pollution snaked out of Asia in April 1998, it flowed over the ocean at a low altitude. Airplanes cross the Pacific at an altitude of six or eight miles—above the usual realm of clouds and weather. This stream of dust traveled at two miles max. And as the river rolled east, the dusts and gases began to react with each other, building complex new dusts.

Fairly quickly, the gases born of fuel combustion condensed into little beads. The sulfur and nitrogen gases, for instance, each sought out their kin and joined together to form supersmall balls—just a ten-thousandth of a hair's width in diameter. If water vapor were scant and the air were dry, these balls may have taken solid form. But the sky usually holds enough liquid to keep them wet.

Some of these pollution balls may have merged with other particles. Nitrogen oxide gas shows a fondness for desert dust. Sulfur beads stick readily to soot grains. They also attract water vapor, creating microscopic droplets of sulfuric acid.

The primitive merged with the primitive, building some very cosmopolitan particles. Bits of mercury from coal fires and carbon from car engines, pesticides from eroding farms, bits of everything from the pollution blanket, all sought a partner, or a thousand partners. The smoke particles and all the pollutants that stick to them combined with their own kind and grew into bigger specks.

The growth of the new-formed dust was eventually stymied. Long before the pollution beads achieved the admirable size of desert dust, they lost the ability to combine with each other. Too big and slow to make more mergers, but too small to settle into the Pacific, they accumulated. And they rolled onward, toward Seattle.

It wasn't the first time, surely, that a river of dust had snaked clear across the Pacific. That must have been happening since there were deserts in Asia. But this time the river was so big and so brazen that people noticed it. Meteorologist Douglas Westphal was so impressed by the dust delivery that he spent some time analyzing its path. At the Naval Research Laboratory in Monterey, California, he had written a computer model that can re-create storms and their paths across the face of the Earth.

"I think we're getting bombarded on a regular basis," he concludes. Slight and dry-humored behind wire-rimmed glasses, he says he missed the whole event. "I didn't see it," he says peevishly. "Oh, if you'd been looking up, you'd have seen it. I guess I was reading."

When he turned his attention to the conditions that shot dust as far and fast as it was propelled in April 1998, Westphal found a rather simple mechanism. South of the dust river, near Hawaii, a high-pressure system had set the atmosphere spinning clockwise, like a big gear. North of the dust river a low-pressure system had set a second gear spinning the opposite direction. The river of dust got its tongue stuck between these two gears and was dragged in. In the movie Westphal's computer model created, the northern gear becomes particularly evident as the second pulse of dust crosses the Pacific.

As the storm begins, a bright cloud of dust swirls in the Gobi. By the following day a tongue has arched up over Russia and struck out across the Pacific. By day three it has straightened and is streaking straight toward northern

Washington. A fresh hot spot grows in the Gobi, and the waning river revives. But this time a gear-shaped spiral of dust takes shape in the northern Pacific. Two days later this sprawling, spinning cloud obliterates Alaska and crosses high above Cheeka Peak. Over the next five days, the thinning river undulates across nearly every state in the Union.

Scientists are elated to have found this ancient river. But they're still missing some pieces of the puzzle, Westphal concedes. Some unknown meteorological event had to boost the dust high enough to catch the fast wind, he says. Equally puzzling, the dust arrived in North America at a *variety* of elevations.

"Some of the dust that came into Southern California was in the middle or low troposphere," Westphal says. "But then in Utah it was at eight kilometers [five miles]. That's cold enough to cause condensation of moisture in the air mass." The dust should have been cleared by falling rain. "We're perplexed," Westphal concludes. "We can't say how the heck it stayed up there."

Although the specifics of that dust shipment remain obscure, Westphal now thinks Asia's springtime dust regularly crosses the Pacific when the wind gears are in place. And when they're not, the dust rivers may peter out halfway, as dust settles into the ocean. Or they may often wend their way farther south, shooting for Hawaii, where Asian dust has rolled overhead so regularly that it no longer turns heads.

Just four days after the first pulse of dust surged out of the Gobi, its tongue flickered over Seattle. At least some of the dust was coming down from its long ride. As the first faint wave rolled over the Olympic Peninsula, it was too high to trickle through Jaffe's instruments on Cheeka Peak, which perch 1,500 feet above sea level. It took another day for the dust to mix down, curl under itself, and pass through the machines heading west again.

And the dust continued to roll in, as the second Gobi storm refueled the river. Day after day the footloose rock continued to call on the coast of North America, from British Columbia to Southern California. The gear-driven winds that had raced across the Pacific were now sinking. And they took their load of dust and little beads of gases and metals, pesticides and soot, on a tour of city streets, greening valleys, and icy mountains.

As they skimmed over the ground, the dusty winds registered their arrival. On April 26, 1998, dust-monitoring instruments all over the West Coast heralded the arrival of the stronger second pulse. In Washington and Oregon the dust levels suddenly hit two-thirds of the federal limit set to protect human lungs. In Vancouver the amount of dust in the air may have doubled. All the way down the coast the story was the same, as dust monitors tallied up concentrations that one might expect to find in a dirty city.

"In Seattle we had one of the worst days of the year," Jaffe recalls. To the casual observer the Asian dust that swirled over the ground was too dilute to be detectable. That is why scientists monitor it with machines: by the time you can see dust in the air, you're breathing terrific quantities of particles. But Jaffe is no casual observer. He remembers biking to work and noticing that the visibility across mile-wide Lake Washington was poor.

The dust that flowed overhead was more striking. People across California, Washington, and British Columbia commented on the eerie, milky sky. Pilots flying over the West reported it. Dan Jaffe stuck a name on it. The phenomenon now suspected of making regular dust deliveries to the Pacific Northwest is called the "Asian Express."

For Jaffe the roar of the Asian Express confirmed a rumor his instruments had murmured a year earlier. In the spring of 1997 Jaffe began monitoring the spring winds as they rolled in off the Pacific and crossed Cheeka Peak. And the instruments immediately began to register pulses of dirty air. Usually, when the team traced those air parcels back in time, they found that the prevailing wind had shifted temporarily and was delivering North American pollution. But sometimes the dirty wind couldn't be tracked back to North America. To be precise, on at least seven days during that first year of monitoring, dirty air had arrived from the west.

Jaffe and others were accustomed to thinking that Pacific air arriving in Washington was some of the cleanest air in the Northern Hemisphere. After all, even a parcel of air that departs Asia heavily freighted with filth has a week or more to shed its dust into the ocean before it hits the Americas.

But in 1997 the instruments at Cheeka Peak hummed out pollution measurements that raised Jaffe's excitable eyebrows. The freshness of the pollution was striking, he noted. If it had indeed come from Asia, it had suffered very little dilution en route.

Jaffe, sitting in his office at the University of Washington, mulls over the politics of globe-trotting dust. The calendar on his wall features Dr. Seuss's *Lorax*, a children's story about the perils of tree snipping. Jaffe teaches environmental science. He knows that the effects of snipping a tree only *begin* to show up when the tree falls. Environmental damage tends to ripple.

"We think this is important in terms of the big picture of how one continent impacts another," he says carefully. "This stuff's gotta rain down somewhere," he adds. His jittery fingers fidget on his knee. "Almost certainly there was mercury in that March event," he adds, raising his eyebrows. "That can get into food chains." His fingers fidget more insistently.

"This is not about how Asia is polluting the U.S.," he continues. "*We* ship our pollution out over the Atlantic, and pass it to Europe. There is no 'away.' Everybody's garbage goes somewhere else."

But Asia is a monumental source of dust and pollutants, and it is growing fast. During the 1999 flights, Jaffe measured incoming ozone gas—a lung-damaging pollutant when it forms near the ground—so thick that it matched the proposed legal limit in the United States. (The EPA's new, tighter limit on ozone pollution is currently bogged in red tape.) This Asian ozone was flying high over American lungs, Jaffe notes, and it may have been a small pollution pulse, not a long-lasting stream: to truly violate the proposed limit, ozone must remain thick for eight hours. But it was a strong hint about how the ancient dust rivers in the sky are evolving.

In April 1998 the Asian Express threw a cascade of dry dust down on the Pacific states. But a thinning river persisted above. The strange, milky sky was reported in Idaho, Utah, and Texas. And then the dust was diluted to invisibility. The dust may have been out of sight, but it was not out of action. This dust was going to make some rain.

As the thinning veil filtered into the Midwest in early May 1998, it drifted over warming earth, perhaps a soggy field of Iowa corn stubble. Simultaneously, the strengthening sunlight drove water vapor up from the land. For every mile this vapor climbed, it cooled by about fourteen degrees. It was soon ready to condense—but on what?

Steve Warren is a lithe and bemused atmospheric scientist at the University of Washington in Seattle. At a picnic table in his flowering backyard he leans over a yellow notepad and draws neat chains of numbers describing the trials and triumphs of suspended water vapor.

The initial problem, he says, is that two water molecules have an extremely hard time sticking to each other. They simply cannot join hands until they are mashed together, cheek to cheek. We think of rain or fog forming when relative humidity reaches 100 percent. But that happens only because clouds are generously supplied with particles for water molecules to condense upon. Atmospheric scientists call these particles "condensation nuclei."

"Without a condensation nucleus, *three hundred percent* relative humidity is required to form droplets," Warren says. If you walked into such a concentrated cloud, you'd be drenched instantly as the vapor condensed on the nucleus of your body.

For water vapor rising into the sky, a little sulfur bead is among the choicest nuclei. From the perspective of a molecule of water vapor, a sulfur bead (a "sulfate aerosol" to scientists) is not so small. Warren's pen flicks out a string of numbers: "A small sulfate condensation nucleus would have about a hundred thousand atoms," he says with a little smile. "Or about a hundred million atoms for the larger nuclei."

High above the Iowa cornfield a drifting sulfur bead suddenly found itself in a crush of rising water molecules. The same fate befell all the other sulfur beads in the Asian cloud: in minutes each was swamped by millions of condensing water molecules. Inside each growing droplet the sulfur bead dissolved. Warren scribbles a formula. When a droplet is one-tenth of a hair wide, he says, the water it contains outweighs the original sulfur nucleus by a million times. And the water vapor also swarmed onto particles of soot and bacteria, the fungal spores and metal dusts, the flecks of fossil and the specks of sea salt that enriched the dust river.

A few minutes prior the rising water vapor was invisible. But now the developing droplets scattered sunlight in all directions. A white cloud appeared in the sky. Each water molecule that condensed onto the newfound dust released a burst of heat into the air. So even as tiny droplets formed, their heat lifted the newborn cloud higher. Only when the whole cloud had cooled to the temperature of the surrounding air did its climb end.

If the surrounding air had been dry and the sunlight strong, these Asia-inspired droplets would have been short-lived. Their water molecules would have hopped off again and dissipated, leaving the original sulfur molecules to resume their bead form. The cloud would have evaporated. This fate is easy to observe in "contrails," the slim clouds that jets leave in their wake. Contrails form when the moist, hot exhaust from an engine hits very cold air. The moisture condenses on bits of soot and sulfur in the exhaust, then freezes. As the heavy ice crystals fall, they shed water molecules back into the dry air, until there's nothing left but the nucleus on which they first condensed. And the same fate awaits most natural clouds, too: only a minority of them live long enough to produce rain that will reach the ground.

This Iowa cloud, however, persists. It ambles east with the wind. Its water droplets begin to merge with each other. About an hour after the collision of water vapor and sulfur beads, some of the raindrops are big enough to fall. A little bit of Asia tumbles down on North America.

And the desert dust in the Asian Express? In this case the earthen dusts will wash out when they're caught up in a falling raindrop. This "scavenging" behavior can deliver colorful precipitation to the ground. The yellow rains that fall downwind of a Gobi dust storm get their color this way, as each falling drop sweeps up oodles of dust specks. Similar desert-tinted rains have been called "blood rains" in Germany, "brown rains" in Arctic Canada, and "yellow snow" in Scandinavia. Unless a fungal explosion ensues, rain really does wash the air clean.

But if the Asian Express had stumbled across water vapor at a higher, colder altitude, it would have been the desert dust, not the sulfur beads, that attracted water. Warm water prefers to condense on a soluble nucleus. But if those little droplets of water are then carried higher into the cold sky, they'll stubbornly refuse to freeze until they find something substantial. As many as half the raindrops that strike the Earth begin their fall as ice, wrapped around a speck of desert dust. Cirrus clouds, which usually form in high, cold air, are populated with ice crystals. But low stratus clouds also can hold ice, as can towering thunderclouds.

"It's not easy to freeze a cloud droplet," Warren says. "You can force it to freeze at about minus forty." But give it something hard to stick to and it will solidify at just a few degrees below freezing. On the yellow pad Warren draws a bucket of water, then draws an ice crystal on the edge of the bucket. "Water in a bucket freezes from the cold edges in." He draws a water droplet with a chilled nugget of desert dust in its center. This time the first ice crystal ventures outward from the dust.

Even if the Asian Express had dived low to the ground before encountering water vapor, it would still have inspired the formation of water droplets. Clouds that form at the surface of the Earth are called fog, but they're really more of the same: various dusts, wrapped in water.

And if the Asian Express had been superrich in sulfur beads, and if it had swung low through a town, it might have turned deadly. The most famous "killing fog" formed in the winter of 1952, when a blanket of warm air descended like a lid over London, England. This trapped cold air near the ground. Sulfurous soft coal was a popular fuel in London at the time. And with no wind to sweep them out, soot and sulfur from the coal accumulated under the lid. As the cold, damp air condensed around the swarming sulfur particles, a thick fog cut visibility in the city to mere feet. Four thousand people died from inhaling the acidic droplets.

Even a polluted fog that forms a bit higher in the air can be lethal. Joseph Prospero is a veteran dust researcher at the University of Miami who usually concentrates his efforts on the migration of Saharan dust to the Caribbean. But when heavy haze off Cape Cod was blamed for the death of John F. Kennedy Jr., Prospero felt compelled to point out the role that heavily polluted dust rivers play in modern weather.

"All the news reports talked about the haze that night and how it was caused by heat and humidity," Prospero told the *Miami Herald* after the mid-July crash. "But in the eastern states, haze is almost always caused by very high concentrations of pollution particles."

A superthick stream of sulfur pollution had drifted east from the coal-powered Midwest on July 16, 1999, Prospero said. Air-pollution monitors and satellite data support this assertion. As the stream flowed past Cape Cod and out to sea; water vapor had condensed on the sulfur beads, forming a heavy layer of tiny water droplets. And once Kennedy descended into that layer of droplets, he apparently could not find his way out.

"We tend to think of decreased visibility only in terms of a loss in aesthetic values," Prospero told the *Miami Herald*. "The Kennedy tragedy shows that there is another cost: Pollution can kill."

Pollution and natural dusts together assure that each part of the world receives its own flavor of rain: Rain falling downwind of the coal-smoky Midwest delivers sulfuric and nitric acids and mercury. Rain falling downwind of a forest fire will drag down soot and tarlike chemicals. Rain downwind from farmland might be rich in soil, fungi, pollen, and pesticides. Downwind from the ocean is the realm of salty rain. Downwind from the great Asian deserts and cities, pollution-scented desert dust falls on the Pacific Northwest and the heartland beyond.

The Asian Express is the dust river of the hour. Ancient, but only recently discovered, it assures scientists of the terrific mobility of dust and pollution. However, additional dust pathways are winding all across the planet. Some of these paths are so heavily beaten that scientists mapped them decades ago. Surely some others have yet to be discovered. But the addition of industrial chemicals to the dust rivers is energizing the search for these subtle skyways. The Saharan Dust Layer, a river that flows from Africa to the Americas, is the most famous of the gritty rivers. And scientists have watched it just long enough to know that it's changing.

The Saharan Dust Layer is so huge that it's hard to miss. Sailors a century and a half ago remarked on the dust that sometimes settled on their ships thousands of miles west of the African coast. But African dust wasn't discovered in this side of the Atlantic until a few decades ago. In the late 1960s, on the Caribbean island of Barbados, a team of researchers had strung monofilament nets on a cliff-top tower, hoping to trap falling *space* dust.

"It was considered by one of the authors to be an ideal site for the collection of cosmic dust, as it was initially thought that little wind-borne land dust could be transported across some [three thousand miles] of Atlantic Ocean," the dust hunters reported in a science journal in 1967. Soon, though, the investigators realized that "the enormous quantity of reddish-brown dust being caught" wasn't cosmic. Thinking it might be coral dust from nearby reefs, they tried to dissolve the stuff with hydrochloric acid. No such luck. They were forced to conclude they had snared the Sahara.

Incidentally, these dust hunters snared lots of other things, too. They caught "cokey balls," a large, porous carbon dust belched by ships burning crude oil. They reeled in diatoms galore, both freshwater and marine species. They bagged dumbfounding quantities of hyphae, the fibrous part of fungi. And they caught white, orange, and yellow "waxy looking lumps," whose origins could not be determined even with heroic efforts in chemistry.

Migratory Saharan dust was soon a subject of intense interest to a small number of scientists, and astronauts began to report huge dust storms they could watch from space. "In the African deserts you see dust fronts as a tongue," recalls Jay Apt, a former astronaut and coeditor of the space-photography book *Orbit.* "I've seen them far out into the Atlantic. Saharan dust looks a bit orange. Mongolian dust is more of a dirty yellow." Over the course of four space shuttle flights, Apt saw just a few major storms. But he caught numerous small, "New Jersey–size" dust storms whirling over the planet.

Down on Earth, dust scholars began tracking the fallout of big Saharan storms. On close inspection Saharan dust was discovered settling on Europe. Then it was found drifting down on Miami and various Caribbean islands. It turned up in South America. And in 1993 a network of dust-monitoring machines tracked a wave of Saharan dust that washed as far west as Texas and north clear to Maine.

This wave first whispered through dust instruments in the Virgin Islands, in the eastern Caribbean, on June 19 of that year. Suddenly, out of a network of about seventy dust monitors mounted in national parks, the one in the Virgin Islands—the only site surrounded by water—was the single dustiest.

By June 23 the wave had pushed north, brushing through a monitor in South Florida. By June 26 the wave had come ashore in Alabama, Mississippi, and Louisiana. On June 30 it covered most of Texas, and it surged north to Illinois. For the next week the wave slid northeast, until it covered the entire Eastern Seaboard.

Fourteen days after the river first registered its presence in the Virgin Islands, the thickest patch of dust lingered over North Carolina, Tennessee, Kentucky, and the Virginias. Thinner dust, still surprisingly dense, was still moving through the Dust Bowl states and the Midwest. As the dust gradually circled out over the unmonitored Atlantic, its trail went cold.

To ensure that the dust monitors weren't picking up a coincidental rash of local dust storms, scientists analyzed samples of the dust. Every major desert has a geological personality that's distinct. One might be calcium-rich, while the next is distinguished by its phosphorous. So when the dust touched down in June 1993, scientists were reassured to find that the chemical character of the dust caught throughout the East matched that of the dust caught in the Virgin Islands.

If the Asian Express was the wake-up call to the Pacific states, the 1993 Saharan dust wave did the same for the East Coast states. That summer in the East, desert-dust readings blew away anything the dusty western states could come up with.

And, like the Asian Express in 1998, the Saharan Dust Layer in 1993 delivered concentrations of dust that made an impressive assault on the legal dust limit set by the federal government. If New Orleans or Jackson, Little Rock or Nashville had already been having a dusty day when the Saharan tide rolled in, the desert dust might have pushed them into violation territory.

And scientists now know that the Saharan Dust Layer is a regular visitor. The eastern United States gets dusted three times a summer, on average, for about ten days per invasion. And a few more times each summer, smaller dust waves powder Florida alone. In Puerto Rico the National Weather Service now issues air-quality alerts when the haze rolls in from the Sahara.

The Sahara doesn't always send its dust west. In winter the Saharan Dust Layer swings southward, on a heading for the Caribbean and South America. Instead of surging up across the eastern United States, the winter waves roll ashore in the Amazon Basin.

And that's only the Sahara's westbound dust. That desert also sends an estimated 100 million tons of dust into the Arabian Sea each year. It ships millions more tons north across the Mediterranean and up through Europe. The Sahara has tumbled from the sky, in the form of gray-orange snow, over northern Scandinavia. Analysis of one Scandinavian delivery concluded that about fifty thousand tons of dust, mainly quartz speckled with rusting iron, traveled more than four thousand miles before snowing out over northern Sweden and Finland. On its long trip north, the clean Saharan dust had gathered toxic, man-made pollutants, which it also delivered to the Scandinavian soil.

Even the dry regions of the United States can sponsor a respectable dust river. Dust from the Mojave makes it offshore of California. During the Dust

Bowl years, U.S. dust sometimes rained generously into the Atlantic. Even as recently as the 1970s a series of dust storms in Colorado, Texas, and New Mexico sent a river of dust eastward over Georgia and into the sea. The island of Bermuda, nearly a thousand miles offshore, may have felt the tickle of this settling dust.

The massive deserts of Australia, though they're old and swept fairly clean of dust, still manage to throw an annual cloud east into the Pacific. The Arabian deserts sprinkle the Indian Ocean. And smaller patches of dry earth the world over produce their own little versions of the dust highway.

All these dust rivers are venerable old things. Blazed countless eons ago, they've moved innumerable tons of desert dust and tiny bits of plants and animals. They've performed their rain- and snowmaking duties faithfully.

But they're changing now. They're growing, for one thing. Joe Prospero, the Miami University dust scholar who has been monitoring the Saharan Dust Layer since the 1960s, has proof. Drought and land abuse in Africa have opened huge new swaths of soil to the wind. The Saharan Dust Layer, says Prospero, has been extra thick ever since the drought began twenty-five years ago.

The Chinese deserts are also spreading. In northern China in the 1950s a political craving to collectivize herders put a stop to the nomadic lifestyle. With animals now trampling and grazing the same patch of ground year in and year out, as much as three-quarters of the Chinese pastureland may be permanently degraded: more easily eroded and less fertile. Agriculture has taken a particularly deep bite out of the Chinese Loess Plateau. These days it's typical for one square foot of the plateau to shed about two pounds of its ancient desert dust each year. The most damaged areas lose closer to six pounds per square foot. Each year the Huang Ho, or Yellow River, carries away more than 1.5 billion tons of *huang tu,* the yellow loess. And the wind carries an untold quantity into the sky.

Even after the lessons of the Dust Bowl, the United States continues to open delicate loess soils to the wind. In 1996 a single, powerful windstorm in Kansas is thought to have beaten an astonishing 650 tons of soil out of each acre it harried—or about 30 pounds per square foot.

As the dust rivers grow heavier, they're also growing more diverse. They are acquiring tributaries of industrial dusts: toxic metals, poisonous pesticides, noxious gases, and a host of other pollutants now travel with the desert dust. These dust rivers never acknowledged national boundaries when they held natural desert dust. And they won't start now.

. . .

Health researchers increasingly fear that the smallest dusts are killing thousands of people each year in the United States alone. So it shouldn't be surprising if some evening down the road the six o'clock news includes a dust forecast along with the pollen count and the thunderstorm warning.

Already, scientists are perfecting computerized "dust models" similar to the ones meteorologists use to predict the behavior of weather systems. Physicist Bill Collins, with the National Center for Atmospheric Research in Boulder, has used one of these dust-forecasting systems to help colleagues locate dust clouds over the Indian Ocean. As the research team searched the air for desert dust and pollutants, Collins was able make predictions about where the dusts of India and Southeast Asia would show up.

"It worked very well," he says happily. "It was certainly adequate for locating large aerosol plumes coming off India."

Buried deep in the computer are estimates of how much fossil fuel each nation consumes, says Collins. That information in turn produces estimates of how much dust and gas each nation releases to the air. Working with these assumptions, the computer predicts how the chemicals will evolve as they head out over the ocean. This information is paired with a weather forecast. Thus armed, Collins can predict not only where a plume of dust is headed but also what kinds of dust and gas it will contain. So far the dust forecast isn't detailed enough to make prognostications for one city, or even one state.

"I've thought about that," Collins says. "If you wanted to apply it to health concerns, you could. But we've been concentrating on international implications. The Asian nations have a significant impact on the climate."

Yes, although the dizzying standard of living in the United States has traditionally meant that we launched the most impressive cloud of fossil-fuel exhaust, our dusty title is now in jeopardy. As developing nations race to achieve a comparable living standard, they're heading for world dominance in the combustion category. As impressive as China's efforts are, India, wielding a similar population, is also achieving dust greatness. That nation's exports of air pollutants have tripled in the past couple of decades. And in recent years wildly acidic fogs have begun to blanket parts of India and Pakistan in winter.

Since developing nations are, almost by definition, impoverished, it is in very poor taste for rich nations to criticize them for attempting to consume a fraction of the energy and material goods that rich nations devour. The scientific approach is to rise above the politics and focus on the effects.

"This is not Westerners going over to browbeat Asia," Collins is quick to note. "These countries want to know what's going on. They're participating."

And it seems inevitable that every nation in the world will need to confront the problem of wandering dust. It will become increasingly clear, as the years pass, that the worldwide forecast for dust is increasingly opaque.

The Asian Express that roared out of China in late April 1998 was breaking apart by the time it roiled across the central United States in early May. The skies gradually cleared. But not for long. Rudolf Husar, director of the Center for Air Pollution Impact and Trend Analysis at Washington University in St. Louis, had quickly mounted an Internet page where scientists could discuss the Asian dust. And on May 9 Husar was inspired to contribute this observation:

"May 9, 1998, was a really 'bad aerosol day' over North America. The Asian smoke pall has arrived to Canada, just north of Vancouver. More Asian smoke is approaching the Pacific coast. The Yucatán/Guatemala fires are still raging, and a thick smoke from there has drifted over the southwestern U.S. A smoke layer has also blanketed much of eastern Canada from fires east of the Canadian Rockies. What kind of neighborhood is this, anyway?"

7

DID DUST DO IN THE ICE AGE?

Twenty thousand years ago the broad, gritty snout of a glacier loomed over northern Montana. The same cold nose of ice sagged south to Illinois, and it blanketed the Great Lakes and the Northeast. The Earth was hitting the bottom of a cold spell. The northern ice cap, fed by endless volleys of snow and ice, had crept ponderously south. Canada was undetectable, and Greenland was a white hump.

It was a chilly era, but hardly the end of the world. Taking advantage of the sunken sea level, fur-clad people may have been scouting a muddy path from what is now Russia to what is now Alaska. Races of enormous animals wandered the open land south of the icy wall. Mammoths and mastodons, huge bears, giant ground sloths, and saber-toothed cats roamed what is now the United States.

Streams of meltwater that rushed from beneath the ice sheet were thick with the powdered bedrock of Canada and the northern United States. The streams dumped their load on dust bars in broad, flat channels. Every time the wind sprang up, more of this glacial dust flew off the braided outwash plain to hang in the air over the glacier itself.

Twenty thousand years ago dust was thick. And although the glaciers were heavy, too—the biggest they had been in the entire ice age—they were about to beat a furiously fast retreat.

Pierre Biscaye stands in his ice freezer at Columbia University's Lamont-Doherty Earth Observatory, on the Hudson River in New York. It's twenty degrees below zero, a cold that creeps steadily up the fingers and solidifies the ink inside a pen. "I've spent hours and hours in here, with only the benefits of my own thoughts," Biscaye says, laughing. He's a burly man, with a grayish beard, brownish hair, and sharp blue eyes. He came to work directly from a racquet-ball game, and for some reason he's not wearing any socks.

His breath comes in white puffs as he draws a butcher knife out of a bag made from a pant leg sewn shut at one end. He lays the blade against a stick of gleaming ice, and chips fly, carrying a history of global climate change, written in dust.

When high-flying dust falls to the ground, it usually blends in with the soil or the sea sediment it settles upon. But when dust falls on a glacier, it is pre-served like flowers pressed between pages of a book. Year after year the snow accumulates, each new snowfall compressing the last.

Biscaye holds the two-foot-long section of Antarctic ice up to a green lamp on a plywood workbench. "See the layers?" he asks. In the bright light the lay-ers are faint stripes, an inch or so thick. To Biscaye they are volumes of infor-mation, stacked in an ice library.

The long cardboard cartons that line the freezer walls hold more plastic-wrapped ice. These cylinders were bored out of deep ice sheets in Antarctica and Greenland with the use of special drills. Eventually, each cylinder will be shaved clean of outer contaminants and will be carried into Biscaye's warm laboratory to spill its secrets.

Although Biscaye and other colleagues have learned to read extraordinary information from the icy pages, big gaps still stymie their understanding of how the world's climate operates. With the global thermometer inching up-ward, they're racing to discover what is pushing the mercury around. Dust is becoming a favorite suspect. They have found that dust occasionally thick-ened in the air by a factor of ten, some say fifty, during ice ages. Just before the end of the last ice age, in fact, dust went through the roof. Scientists aren't dis-missing this as a coincidence. They want to know what, exactly, dust's role was in the glaciers' demise.

One very *clear* message in the ice is that the Earth's climate is naturally erratic. According to the dust and gases trapped in the ice, the climate is always—al-ways—in flux. If it's not getting warmer, it's getting colder. Year to year the shifts may be masked by an El Niño, a La Niña, a Mount Pinatubo, or some other temporary drama. But decade to decade, century to century, the world's temperature is in constant motion.

On a grand scale our moderate, modern climate is abnormal. Through most of the dinosaur era the planet's normal state was decidedly steamier.

When the *Oviraptor* perished in the Gobi Desert, the world may have been eleven to fourteen degrees hotter, on average.

Then, just 2.5 million years ago, the planet entered a pattern of periodic ice ages, punctuated by brief warm spells. The ice caps, as a result, have taken to advancing and retreating intermittently. The glaciers have ruled for the lion's share of time, with the warm "interglacials" lasting roughly ten thousand years each. We inhabit an interglacial known as the Holocene, which ought to be coming to an end any day now. The thermometer, however, does not seem poised for a plunge.

All things being equal, no climatologist would be surprised if the Holocene persisted for another few thousand years—climate change is that erratic. But all things are *not* equal. Human industry has wrought profound changes in the Earth's atmosphere since the last warm period.

Based on "fossil air" trapped in the same ice cores that preserve dust, scientists can show that industrial-era humanity has added almost 30 percent more carbon dioxide to the atmosphere than was naturally present. Carbon dioxide rises into the air whenever fossil fuels or living things burn—or even when they peacefully rot. Unfortunately, carbon dioxide is not one of those drifting gases that condenses on dust and falls back to Earth. It accumulates in the air. And there it traps heat that would normally radiate off the planet and into space.

In the past hundred years the planet's average temperature has risen by a degree, more or less. Maybe this sounds like a small change. But it is the same size shift—in the opposite direction—that brought on the "Little Ice Age" of 1450 to 1890, a comparably small global dip that caused European rivers to freeze, glaciers to advance, and precipitation patterns to migrate. A little shift has a big impact.

But carbon dioxide isn't the only item pushing the temperature around. Everything that floats in the air has an effect on how much sunlight reaches the Earth. A sailing grain of pollen, for instance, may absorb some heat. A spider leg might reflect some light. And how well do flying fungi attract the water vapor that makes clouds? What is the net effect of a billion tons, or 3 billion tons, of flying desert dust? The air is filled with an array of dusts. But determining how each breed of dust effects the climate—that's a work in progress.

The Earth's erratic climate has become the subject of many computer models that incorporate information about winds, sunshine, clouds, and other forces that affect the temperature. Some of these forces, such as carbon dioxide gas, are "positive," pushing the mercury up. Others, such as a reflective blanket of snow, are "negative," pushing the mercury down. But the computer models are only as good as the information they're fed. And for many dusts, good information simply doesn't exist yet.

"The climate modelers know [desert] dust is important," says Pierre Biscaye, "but it's the *least* well known parameter in the Earth's thermal balance. Right now," he adds, forehead furrowed, "they don't even know its *sign*."

Biscaye is not a modeler. He is that breed of scientist who works on "ground truth." His hands-on work with dust builds a portrait of the past. And his observations help computer modelers to build more realistic models of the future.

He locks the ice freezer and walks through the world's biggest mud collection. Just as drills can cut a core sample from an ice sheet, they can also pull up a plug of sediments from the sea floor. Thanks in part to the special zeal of Lamont-Doherty's founder, Maurice Ewing, the institution is home to a staggering amount of such mud. The entire cavernous room smells of musty clay. The dry, broken cores lie in skinny boxes on metal racks. Much of the mud is pale with the remains of algae. Laid end to end, these cores would stretch more than eleven miles. Many of them have never been analyzed: they're unopened books. But they represent a map of every major ocean and sea in the world. As Biscaye leaves the building and crosses the street to his own office, he says his dusty career began with sea mud. But ice, he now believes, is a better storyteller.

He opens his office door to reveal a small forest of plants hanging under a skylight, dozens of paintings and photographs of his family, curly piles of maps, an aging herd of armchairs wrapped in Indian-print bedspreads, and a stuffed tiger head wearing 1970s-vintage spectacles. He reaches past a collection of bird wings pinned to the wall (all from road kill, he notes) to tap a world map.

In the deep ocean off the coast of South America, he says, desert dust surely must fall. But rivers also dump their sediment into this underwater valley. And deep currents run the length of the continent, stirring together all sorts of sinking dusts.

"Saharan dust carries a particular strontium isotope signal. But if you look for it in there, you've gotta look for it among other stuff. There's Saharan dust in there," he says, rapping the valley. "But you can't see it."

"Now, in an ice core, there is no other way that continental dust can get there, except through the atmosphere." So to learn what forces alter the climate, Biscaye reads the dust caught in ice. The ice is insistent on one fact. And although the meaning of this fact is unclear, it could be of immense import: during glacial periods there is a lot more dust in the air.

On a bulletin board outside Biscaye's office a display of dust data is introduced thus: is the extra dust merely a side effect of a cold climate? Or did this excess dust actually bring on a powerful episode of global warming?

The solution to that riddle has not emerged from the ice. But a few intriguing theories suggest that the billowing dust actually killed off the glaciers.

Something besides carbon dioxide is clearly kicking the thermometer around, says Dean Hegg. A lean, soft-spoken man with snowy hair, Hegg is an atmospheric scientist at the University of Washington in Seattle. He analyzes how individual species of dust and gas push the temperature one way or the other.

"There's this big issue: Computer models say you get a big climate change with carbon dioxide—a much *bigger* change than we actually see," Hegg muses. "But if you add aerosols [particles] to the model, you get better agreement." So is the net effect of all dusts to push the mercury *down*?

Climatologists sometimes talk about the surface of the Earth in terms of the wattage of solar energy it absorbs. If you scrambled the planet's surface into a homogeneous color and texture, each square meter would absorb 240 watts of energy—enough to power a small chandelier. The reason we don't fry is that this energy radiates back to the cosmos, especially in the dark of night. The Earth gives away as much as it gets and therefore maintains a balance. Or it used to.

Now, the extra carbon dioxide, methane, and other gassy by-products of human enterprise trap some of the heat that radiates back toward space. This nudges up the global temperature. And when the computer modelers add the human-made fraction of these gases to the virtual world, the computer predicts that the planet will heat up at between two and six degrees per century.

But, as Dean Hegg says, that rate does not appear on the thermometer. Although the surface of the planet is warming, the rate is less fiery than what the computer models predict. Now the computer modelers are slowly feeding dusts to their models, as data become available, to see if they can find the missing "coolant."

Hegg is one of the people trying to find the most potent dusts. He has investigated sulfur beads, tree-goo beads, and desert dust, in addition to plain old water vapor.

"Simple water vapor may be a potent forcer," Hegg ventures. "And we know that sulfate aerosol plays a large role—and there's a lot of it. Then, a lot of the aerosol in the atmosphere is organic—both anthropogenic and from trees—and the role of these organics in forcing is underestimated.

"And there's [desert] dust. It's very tricky. And, again, a lot of people feel we've underestimated the amount of [desert] dust in the atmosphere."

But even if scientists knew whether a grain of soot or desert dust will warm or chill the climate, they still lack of good census of *how much* of each dust rambles overhead. And so Hegg and others are left guessing at the total impact of each breed of particle.

For now let's consider how some of nature's dusts affect the planet—to the extent that this is understood by scientists. Then, we'll consider our human contribution.

The most obvious trick of sailing desert dust is to interfere with sunlight. In March 2000 a monster dust cloud left the Sahara and swirled over the Atlantic Ocean. Though the cloud's size, captured by a satellite camera, was jaw-dropping, its most striking feature was how sharply its gold color contrasted with the dark Atlantic. It was a classic demonstration of an old truism: Dark colors absorb, light colors reflect. And in this case each sunbeam the dust turned back represented a small heat loss for the planet. So does desert dust have a natural cooling effect?

Don't carve that in stone. Dust is more reflective than many parts of the Earth's surface, including dark oceans, forests, and mountains. But compared to snow, or clouds, and even some bright deserts, Saharan dust is actually darker. Thus, when a plume of flying desert dust spreads above a patch of superreflective ice, or clouds, or certain deserts, it actually *reduces* the amount of sunlight that the planet bounces back to space. Furthermore, as desert dust blows across the sky, it absorbs some of the sunlight that strikes it. So dust is a natural warmer, too. This is why it's hard to pin a sign on some dusts: there are many factors at work simultaneously. All things considered, concludes one climate-and-dust expert, Ina Tegen, desert dust's most dependable effect may be to shift heat from the Earth's surface to its gaseous atmosphere.

As for the other sorts of dust that Nature sends into the sky? Each item has a unique relationship with sunlight. Dean Hegg has analyzed the air over the Atlantic and come up with a rough ranking of which airborne particles most ably intercept sunlight. Tiny droplets of water blocked the most. But the second-strongest sunlight squelcher proved to be carbon-rich organic dusts, that oily family of compounds released by fires, factories, and trees alike. Next was dry sulfur beads.

This study, however, addressed just one piece of the global patchwork—the Atlantic Ocean. Other researchers have proposed that sea salt wields the biggest force over remote parts of the ocean and that desert dust is the mightiest forcer over the entirety of the North Atlantic. Ina Tegen, who uses computer models to predict the effects of dust, has proposed that when averaged worldwide, sulfur beads, dust, and carbon-rich particles all intercept about the same amount of sunlight.

But blocking sunlight is just the first of dust's tricks. Natural dusts in the air exert an *indirect* force on the climate by tampering with clouds.

Water vapor, recall, has terrific difficulty condensing without dust to cling to. So if there were less dust in the sky, the bright reflective clouds that swirl over the planet would be rarer. And if there were *more* dust, clouds would blanket more of the planet.

Clouds usually cloak about half the Earth. Their presence doubles the planet's natural reflectivity. And that's a lot of heat bounced away to space. Clouds, then, are a powerful "negative force," cooling the climate.

But, to keep climatologists tugging at their hair, clouds can also be *positive* force. Anyone who has spent a winter in the chilly north knows that a cloudy night is a blessing: a blanket of clouds overhead slows the Earth's heat in its nightly rush toward the stars.

Clouds come and go, form and fade, due to many circumstances. Dust is one of them. Without sufficient dust, water vapor is stuck in limbo. But when there is *extra* dust in the sky, water vapor is divided among so many particles that no one droplet ever grows big enough to fall. A dusty cloud may hold twice as many water droplets as a normal cloud, but each droplet will be half the normal size.

The implications of this small, dusty difference are immense. A NASA satellite that peers into tropical clouds has revealed that the frequent fires of the tropics produce soot—black dust—which can effectively turn off the rain faucet. As this satellite watched clouds crossing the Indonesian island of Kalimantan in 1998, it caught a striking difference between normal clouds and smoked ones.

In a normal cloud a million tiny droplets must coalesce to build one raindrop that's heavy enough to fall. In an extra-dusty cloud, where the water is divided into even smaller droplets, it takes many more droplets to make a successful raindrop. So an extra-dusty cloud takes longer to produce raindrops—if it manages to produce them at all. And sure enough, the NASA satellite showed that clouds crossing the smoke-free sectors of Kalimantan rained freely. But clouds that passed through smoke plumes over the island were unable to shed their water.

Volcanoes, which produce a bonanza of cloud-forming sulfur particles, also seem able to shut off the rain faucet. One study of rainfall in Taiwan found that when volcanoes upwind of the island threw up dust, rainfall on Taiwan dwindled.

The superdusted clouds have other climate implications, too. The smaller, more numerous droplets in the dusty cloud are more reflective than normal-size droplets, and they bounce more sunlight back to space. That chills the Earth below. The difference is sometimes obvious to the naked eye: A superwhite cloud is probably jammed with extra-small droplets. And a dusty cloud lingers in the sky, stretching its climate effects over a longer period. By day it reflects sunlight longer. At night it traps heat longer.

Natural dusts, then, have multiple methods for fooling with the temperature of the planet. The challenge for Pierre Biscaye and other dust sleuths is to find hard evidence for what kinds of dust were cruising which dust rivers in the last ice age.

The first step, says Biscaye, is to determine the original source of the dust that fell on glaciers. Biscaye pokes a small plastic bag of dust tacked to his bulletin board. The dust is finer than flour, and pink-gold. It is labeled WEICHANG, CHINA. Beside it hangs a second bag, holding smoky-brown dust from Eustis, Nebraska. He did not extract these dusts from ice cores. To get this much—a thumb-and-finger pinch—from ice would require the destruction of yards upon yards of precious ice core. These bags hold modern dust, scooped off the face of the Earth. The dust in every desert and glacial-outwash plain on the planet has a unique mineral signature. And Biscaye aims to collect the autograph of every major rock-dust source in the world.

"I've spent a lot of time trying to get these samples," he says. "I've got hundreds. When a friend says, 'I'm going to Siberia,' I say, 'Take some plastic bags!'"

Now when he extracts precious dust from precious ice, he can compare it to the samples in his collection. This is how Biscaye and his colleagues pinned down the source of dust that settled on Greenland during the last ice age.

He leads the way to the room where the book of ice opens its pages. He has just finished vaporizing an ancient section of Greenland ice. A glass-topped canister resembling a large pressure cooker sits on a bench, seemingly empty. Biscaye picks up a flashlight, then flips off the overhead light to show that the canister is not exactly empty. In the beam a clear plastic baffle appears, mounted inside the canister. And on the plastic there is dust. It's a breathtakingly small amount. In fact, one human breath on a cold windowpane would amount to more matter.

The dust was left behind when the ancient ice was sublimated molecule by molecule and sucked away to a separate container. Biscaye can't simply melt the ice and filter out the dust, because some kinds of dust—like gypsum and calcite—might dissolve in the water and flow away. And those can form an important facet of a dust's identity. This faint powder is the product of perhaps five pounds of ice, but it's only one-quarter of the amount Biscaye would need in order to obtain a good signature.

"That's the disadvantage to working with ice cores," he says. "There isn't a heck of a lot of dust. You've gotta work like the dickens to get every grain to make your measurements. At every step of the way you have to fight to keep each grain. Every container you put it into, it sticks to. Including Teflon. When I empty a container, I wrap my finger in Saran Wrap and gently wipe the container again. And I always find more."

To keep ordinary dust in the laboratory air from contaminating the antique samples, Biscaye handles his dust on a special bench cased in glass. Superclean filtered air blows from inside the case. Human hands poke in through a hole, and the superclean wind rushes past them, pushing back any dust that might attempt to invade from the laboratory. To demonstrate the dustlessness of the bench, Biscaye thrusts a handheld particle counter through the hole and allows it to sample a cubic foot of air: a red zero sits stubbornly on the readout. He repeats the measurement in the open laboratory. This is a relatively clean lab, isolated in the woods flanking the Hudson River. Biscaye flicks it off as the readout races past ten thousand specks of dust.

In this lab Biscaye has extracted the dust from Greenland ice dating to the peak of the last ice age—23,000 to 26,000 years ago. And that dust had a bold, decisive signature. On a graph that shows big peaks for plentiful minerals and small peaks for rare ones, the Greenland dust had a black mass of narrow peaks on the left and then two fat, tall peaks on the right.

Back in his cluttered office, Biscaye climbs onto a chair and fetches a globe from atop a filing cabinet. Turning the North Pole up, he stabs Siberia, north of the great Asian deserts.

"The computer models say the dust in Greenland came from a large area of eastern Asia, centered in northern Siberia," he says, frowning. "Well, it may be easiest for a computer model to derive the dust from there," he continues, jabbing Siberia with his finger. "But I don't think that's where the dust came from." He can't eliminate Siberia just yet, because he hasn't finished analyzing new dust samples from that area.

But the Greenland dust itself argues for a different source. Its signature is markedly short on a mineral called smectite. When Biscaye noticed that, he began flipping through his dust collection, looking for a match.

In addition to Siberia, the Sahara had been blamed for dusting Greenland. The Sahara is probably the biggest source of desert dust in the world, after all. And it has a modern-day reputation for occasionally yellowing the snow of Finland, and even Svalbard, well north of the Arctic Circle. Greenland isn't such a leap. But the Saharan signature shows a strong peak for smectite. So the dust is not Saharan.

The generous loess blankets of the midwestern United States had also been charged with dirtying Greenland. However, Biscaye notes, midwestern loess is rather famous for its *abundant* smectite. Cross off North American dust.

Unless the source was rock flour from the glacial-outwash plains of Alaska? In that case Biscaye found that dust samples from Alaska contained decidedly different isotopes of strontium, neodymium, and lead than did the Greenland dust. That told Biscaye the parent rocks of the two dust samples have different ages. And ditto for a dust sample from the Ukraine.

Biscaye holds a printout of the Greenland dust's zigzag signature in one hand. Over it, printed on clear plastic, he lays the zigzag signature of the Chinese Loess Plateau, that heavy blanket of dust blown out of the Gobi and Taklamakan Deserts.

"Boom," says Biscaye.

It is not an exact match. But the boldest peaks and valleys line up. Like the Greenland dust, the Asian dust has a smectite valley. The dusts share a similar signature in strontium, lead, and neodymium isotopes, too.

"Looks like a candidate," he says. He's not claiming certainty. Turning the globe again, he raps his knuckles on western Asia. "When we first published this, someone could always have said, 'Did you test samples from, uh, Tajikistan?' Well, no," he says. "We didn't have any. But *now* we have samples. And we have samples from Siberia, too."

Dust can also tell Biscaye something about the *wind* during the ice age. When Biscaye samples the uppermost layers of Greenland snow—modern snow—he can find the signature of Asian dust. But there's not much of it. And the average grain is tiny.

These days, Biscaye explains, the typical Asian dust storm begins to run out of steam over the Pacific Ocean, thousands of miles short of Greenland. As the dust settles into the ocean, only the finest, lightest grains are left to travel on, across North America to Greenland. But in the layers of Greenland ice from the last ice age, Biscaye finds three to ten times more Asian dust. And some of the grains are bigger than those of today.

Many climate scholars think that winds were stronger during the ice ages.

Instinct would suggest that stronger winds could kick up more Asian desert dust and carry bigger grains of it farther. A second kind of dust in the ice supports this idea. Sea salt preserved in the ice cores rises and falls in concert with the desert dust. This, too, stands to reason: a faster wind would build more whitecaps on the ocean, break more bubbles, and carry off more salt particles.

Another school of thought, however, claims that the dustiness of the ice age is more the work of expanding deserts than strong winds. When the Earth was colder, water evaporated more slowly from oceans and lakes. Rain was rarer. Deserts grew and produced more dust. One computer model suggests that cutting global precipitation in half would result in air ten times dustier than it is today.

Biscaye's dust weighs in on the side of strong winds. The Greenland ice shows a threefold to tenfold increase in dust during the glacial max. If you assume the wind was no stronger than it is today, Biscaye argues, you would require three to ten times more *desert* to produce all the extra dust. Such an enormous expansion, he says, would surely require the erosion of new and different rocks. And that would change the signature of Asian dust. Did it?

No, says Biscaye. Over the three thousand years' worth of ice he has opened, the dust grows thicker and thinner, but its signature stays fairly constant. One mineral that accumulates in northern rocks as they age does make a stronger showing during the ice age. When he adjusts for this, Biscaye is left with a modest northward creep of the Asian dust source during the ice age. The deserts may have shifted a bit, Biscaye concludes, but strong winds were more important.

And so you might write in dust, if not carve in stone, that dust behaved differently in at least two ways during glacial times. Dust rose up from deserts that may have been bigger than they are today. And stronger winds carried that dust farther. But was the extra dust of the ice age up to the task of killing off the glaciers?

To secure additional evidence, Biscaye turns to Antarctic ice. This ice can be more difficult to read, because each year's snowfall is much scantier. The layers are less clear. The dustfall is thinner, too. But whereas Greenland's ice goes back only about a hundred thousand years before reaching bedrock, the book of Antarctic ice may cover half a million years of world history, encompassing four complete ice ages and their warm interglacials.

Collecting ice cores is rarely a stroll in the park. In Greenland it requires camping in shacks atop the glacier for months, as the sections of ice are carefully raised, cataloged, rinsed in purified water, and wrapped in plastic for storage—because the ice can't travel right away.

"You're pulling ice up from huge depths," Biscaye says. "You're releasing enormous pressure. If you try to cut it, it can just shatter. For that matter, it can shatter just coming up the drill hole." And so the cores spend their first year topside resting, slowly relaxing into their new low-pressure lifestyle. Then the cylinders are sent to the National Ice Core Laboratory in Denver, to be quartered lengthwise and distributed to researchers.

Retrieving ice from tropical glaciers in Asia can be even more adventuresome, involving transportation of hard-won ice on the backs of yaks. But Antarctic coring sites are hard to beat for remoteness. The ice of eastern Antarctica, the side nearest Australia, is best known. Like the dust in Greenland ice, the dust in east-Antarctic ice also multiplied during peak glacial times—by as much as tenfold. And as with Greenland dust, computer models of the Earth's climate had their silicon hearts set on a specific source for the dust that fell on east Antarctica in the last ice age. The models, Biscaye says, used to like Australia.

Compared to the Northern Hemisphere, the Southern Hemisphere is desert-poor. At first glance Australia's big, bald interior looks like a great place to find dust. Additional bits of desert, in southern Africa and South America, might also seem tempting. But deserts aren't the only great dustmakers.

"The Andes are young and volcanic," Biscaye says, again tipping the globe on his lap. The fat tail of South America, edged with dark mountains, droops down toward the white disk of Antarctica. "During the last glacial maximum, this was totally ice," he says, dragging a finger along the strip of Patagonian peaks. "The pressure of more and more snow made glaciers. These bore down on the mountains, grinding them and producing huge amounts of rock flour. And at the base of the glaciers there were streams absolutely *choked* with this rock flour." His arms weave through the air, his fingers spread like streams searching for a way around their own burden of silt. "It left *huge* fields of glacial outwash. And once these were dried out, they made a great source of wind-blown dust."

And a second notable dust source challenged Australia's computer-alleged role. During an ice age 12 million cubic miles of seawater may have evaporated, been carried over land and deposited as ice. In the last ice age the glaciers locked up enough water to lower the sea level by about four hundred feet.

Now, what a clam might think of as coastal mud is really just waterlogged dust. Drain off the seawater, bring on the sunshine, and mud will dry to powder. As the glaciers grew and the sea retreated, mile after mile of mud was opened to the wind. Some of this mud must have risen into the air.

So the strong winds of the ice ages had a few dusts to choose from as they spiraled toward the South Pole. If Biscaye wanted to pinpoint the source for the dust that fell on Antarctica, he would have to consult his dust collection.

He and his colleagues fingerprinted dust from Africa's Namib Desert, and loess from the Kalahari. They checked loess from New Zealand's Canterbury Plains and from Australia's Great Sandy Desert. They considered ocean mud from the once exposed continental shelf east of Argentina and more mud from the Falklands. They looked at the soil of Tierra del Fuego at the southern tip of Chile. They examined Patagonian loess—and the signatures matched. One more ancient river of dust was sketched onto the map. And dust sleuths got one step closer to understanding how dust imposes itself on ice.

David Rind has caught a glimpse of dust's malignant side, and despite broad gaps in the ground truth, he's not shy about discussing his vision. A computer modeler for an Earth-oriented division of NASA, he's thin, with a quirky grin and dark hair. Rind says climate simulations suggest that two dustmakers—glaciers and the Asian deserts—could have pumped out enough dust to raise the temperature in their immediate neighborhoods by a blistering nine degrees.

When a layer of dust hangs in the atmosphere, says Rind, the Sun heats it. That hot layer halts the normal circulation pattern that raises moist air off the ground. Without moist air aloft, there will be no rain. Without rain, the ground below will slowly dry. And dry ground will heat up.

For comparison, consider your skin. Human bodies are designed to ooze sweat because evaporation of that moisture cools us. The Earth uses the same method: water evaporates off its surface and carries away heat. But when a patch of ground runs out of "sweat," the temperature starts heading up.

"In the final analysis," Rind concludes "this effect on surface warming of the Earth can be larger than the effect of the dust itself."

This would have worked only over land, Rind says, and only in the areas where dust was thickest. And the dust would have been plenty thick near glaciers that produced it. As the rising dust caused the ground below to warm, the neighboring ice would have begun to soften.

"When glaciers start melting, they produce lakes," Rind says. "Some people think that what might have happened is you get waves on those lakes. And if the tip of a glacier lies on top of a lake, the waves will lift it up and down. So you could get mechanical breaking." And this ice-busting action, the theory goes, would shift glacial breakdown into high gear.

As a reminder, this scenario was cranked out by a computer model, and models are supplied with something less than the zillions of natural variables that actually influence climate. For instance, Rind says that something as subtle as the *color* of the rising dust could nudge the outcome in one direction or another: the dust of dark basalt absorbs more heat than the dust of pale granite. So this scenario is fairly speculative.

But despair not, those who prefer a dark ending to a tale. Another answer to the riddle of the retreating ice may be that the dust churned out by the grinding glaciers actually came back around to bite them *directly:* in this version the windblown dust settles *on* the ice, absorbs sunlight, and warms the ice to death.

Steve Warren, the Seattle ice-and-atmosphere scholar, has toyed with this idea. "When I first got started in atmospheric science twenty years ago, I thought dust was the most boring topic imaginable," he confesses with a grin. "But then, when I investigated its effect on snow, I became fascinated."

When even a tiny amount of dust contaminates fallen snow, Warren says, it increases the amount of sunlight the snow absorbs. Clean, fresh snow reflects 80 percent of the sunlight that strikes it. But dirty snow absorbs more solar energy. The energy is converted to heat, and heat is no friend to snow.

Warren says that each breed of dust affects snow differently. The most powerful is soot, he says. It holds heat about fifty times better than desert dust and two-hundred times better than volcanic ash. But ash is effective, too. He holds up a photograph of Blue Glacier, on Washington's Mount Olympus. In the center is a neat rectangle of snow, about ten feet by thirteen feet, which rises above the surrounding field. In 1980, Mount Saint Helens sprinkled the entire glacier with volcanic ash, Warren explains. Opportunistic researchers marked off this rectangle and scraped it clean of ash. Then they waited. Although volcanic ash is on the weak side as a melting agent, in just two weeks it soaked up enough heat to lower the level of the snow around the ash-free rectangle by about a foot. Dust diminished ice.

A year later, in early March 1991, a heavy fall of Saharan dust produced a different kind of scientific breakthrough. By that summer, which was unusually warm, the dust had melted enough ice to reveal the mummified remains of "Oetzi," the Copper Age hiker who had lain buried with his tools for over five thousand years.

Earth-systems scientist Tamara Ledley took up where Steve Warren left off—but using a more dreadful dust. "I put Steve's work into my computer model, to see what might happen after a nuclear holocaust," says this dark-eyed woman with a broad smile. "What if all that smoke goes into the atmosphere, and then some of it settles on sea ice? Basically, what would happen is the sea ice would melt away."

That was a start. Over the next ten years Ledley slaved over the model, steadily improving it. "I spent an awful lot of time adding other factors," she says. "I added snow to the ice. I added leads—cracks in the ice. I added the interaction with the atmosphere. I refined the method for calculating snowfall rates. I added the movement of the ice itself."

When Ledley eventually departed academia, she passed the baton to Rice University student Robert Steen, whom she continued to advise. Steen turned the computer's attention to great sheets of ice that grew during ice ages. And at the peak of the simulated cold spell, he sprinkled the ice with all the dust that should have been riding the wind at the time. And the ice began to melt.

"If you have little tiny black pebbles on ice," Ledley explains, "the pebbles absorb solar radiation. And then they melt the ice around them. Dust does the same thing. Now, the water runs away, but the dust stays at the surface. More and more dust accumulates on the surface, and it accelerates the process."

Enough to melt away a mile-thick sheet of ice?

"Eventually," says Ledley. "But maybe that process starts the melting, and then other things come into play. Once you lower the surface of the ice, for instance, it's exposed to warmer air, and it melts faster. So dust could be a *trigger*."

Ledley had always wanted to make this case and waited impatiently to see what her student would wring from the computer model. When she finally received his data, she was headed off on a trip. She took it to read enroute. "I saw his graph, and I realized we finally managed to do it!" she remembers. "And I was stuck on an airplane—there was no one I could tell!"

Then the student moved on to other topics, and Ledley was wrapped up in her new work, and this oedipal thesis has never been polished up for publication in a science journal. Scientifically, this data on ice-killing dust qualifies only as a rumor, until it has been critiqued by Ledley's peers and published. But it suggests that a little dust might move a mountain of ice.

Case closed? Of course not. Dust in the air, as we've seen, can both heat and chill the planet. So why should we expect anything less confusing when desert dust settles back to Earth? Yes, when it lands on glaciers, it may heat them treacherously. But recent experiments are hinting that when dust falls in the ocean, then it can indirectly *chill* the planet.

One of the nutrients that ocean plankton need is iron. Without a bountiful supply, they can't pull off a bloom. So less iron means less plankton—which means less of the sulfur-rich chemical DMS (dimethyl sulphide), which rises into the sky when plankton perish. That means fewer natural sulfur beads soaring overhead. A shortage of sulfur beads makes clouds whose few big drops rain out quickly. More sunlight reaches the Earth. Therefore, a death of plankton makes a warmer planet.

Now add iron: More plankton, more DMS, more clouds, less sunlight. Thus, a surplus of plankton makes a cooler planet. As a bonus, growing plankton pull the global-warming gas, carbon dioxide, out of the air.

"You give me half a tanker full of iron; I'll give you another ice age" was the summation of the late John Martin, father of the "iron fertilization theory." Martin believed that dumping iron into the ocean would start an algae bloom and lead to a fresh layer of bright, reflective clouds.

And as far-fetched as it sounds, some small experiments in iron dusting suggest that he was correct: stirring iron into the ocean's surface *does* bring about a plankton bloom. But how would this happen without a tankerful of iron and meddling scientists?

Many of the Earth's rocks are rich in iron. Therefore, many of the Earth's high-flying dusts are iron-fortified. And scientists are taking advantage of this situation to investigate whether iron that tumbles naturally from the sky can inspire a plankton bloom.

One team focused its attention on the northeast Pacific, which has iron-poor waters. Despite the presumed iron shortage, the team found that algae do enjoy periodic blooms there. Many of these clustered in the summer months, researchers noted, which was a vote against the most obvious candidate: airborne Asian dust is most common in spring. But satellite imagery offered another source of iron, in the Copper River Valley of Alaska. In one image a pale stream of glacial dust streaks out of the valley like cigarette smoke, to ripple over the North Pacific.

The dust hunters also noted that Alaska is pockmarked with ash-spitting volcanoes. Though they made no direct observation of ash raining down as plankton food, they did note that a previous study linked Pacific algae blooms to the Mount Pinatubo eruption. What with volcanic ash, Asian deserts, and Alaskan glacial flour, the dust hunters concluded, there is plentiful airborne dust to explain the plankton blooms of the northeast Pacific.

So in the dusty ice age perhaps well-fed plankton produced more sulfur and more bright clouds. Some plankton fans have even proposed that modern plankton blooms could naturally offset global warming. But in the labyrinthine climate system their actual effect on the thermostat is still unclear. For instance, although scientists have demonstrated the link between iron-rich dust and plankton blooms, no one has firmly connected plankton blooms to an ensuing bloom of clouds.

These are all alluring stories: A bumper crop of dust heats the ground, and that melts the glaciers. Or dust falls directly on the ice and melts holes in it. And maybe plankton, tiny wild cards in the climate game, fight the trend, using dust to keep things cool.

But Pierre Biscaye, for one, hasn't seen enough ground truth yet to lean one way or the other. Maybe dust busted the glaciers. Maybe it didn't.

"I don't think anybody has a good explanation," Biscaye says, still shaking his fingers after his trip to the freezer. During the previous ice-drilling season in Greenland his hands became permanently sensitized to cold. "It's a wide-open question."

There is probably some truth to each tale of dust and ice. And the sooner the stories are sorted out, the better. Climate modeler David Rind estimates that the air today holds as much dust as the air just before the ice retreated twenty thousand years ago.

If it *was* dust that sent the ice packing, it certainly worked fast. One century ice sheets ruled the north, and dust clouded the air. Then in the next century temperatures may have shot up five to seven degrees. The ice was on the run, and the dust was on the wane.

We live in precarious times, climate-wise. Judging from past climate swings preserved in ancient ice, the glaciers should soon be creeping southward. Instead, the mercury seems to be chugging upward, and the ice is still melting.

We've already seen how nature's own dusts might push the mercury this way or that. But what are the deeds of humanity's dusts? The Earth's climate system was a baffling and erratic machine before we ever lumbered onto the scene, and now we've emptied Pandora's box into those mysterious gears: a staggering load of sulfur beads and nitrogen-rich bits, boatloads of soot, and a smattering of everything else that we build or burn. We've even added to the desert dust.

Ina Tegen has estimated that fully *half* the desert dust in the air today may rise from land damaged by human use. If that's true, then the planet's desert-dust load is twice what it was in preagricultural times. And what is this extra desert dust doing to today's climate? Is it pushing us toward an ice age? Or a meltdown? As Biscaye says, desert dust is a crucial piece of the machinery, but we haven't figured out whether it's a warming piece or a cooling piece.

Not every dust that humanity throws into the works is so hard to size up. Climatologists have a modest handle on the role of highly reflective sulfur beads. Recall that each square meter of the Earth's surface receives, on average, 240 watts of sunlight. Steve Warren says that the atmosphere's average burden of sulfur particles now chills the Northern Hemisphere by a bit more than one watt on average.

Because the Northern and Southern Hemispheres swap air very slowly, sulfur beads tend to rain out before they afflict the less industrial half of the world. In fact, sulfur beads are such powerful rainmakers that they don't even spread through the Northern Hemisphere before they fall. Their watt-blocking effect probably is concentrated in "sulfate shadows" that spread downwind of the world's industrial areas. For what it's worth, computer models predict three particularly chilly shadows. The first, courtesy of midwestern industry, spreads offshore of the East Coast of the United States. Another cold spot afflicts Central Europe. And a third blots eastern China and a swath of the neighboring Pacific Ocean. Although sulfur beads don't evenly blanket the world the way carbon dioxide does, their chilling effect is so strong that they may nearly cancel out carbon dioxide's warming. This apparently beneficial coincidence led a colleague of Dean Hegg and Steve Warren's at the University of Washington to make a surprising observation in 1999.

"You could reach the frightening conclusion that we have learned how to pollute just right to prevent climatic disaster," wrote atmospheric chemist Robert Charlson in a press release. You could—but you shouldn't. Charlson continued, "The greenhouse effect works twenty-four hours a day, everywhere, and the [sulfur] aerosol effect only works during the day, and only in certain places."

Dean Hegg points out another small problem: if humanity suddenly ceased all polluting activities tomorrow, rain and snow would drag most of our cooling sulfur beads out of the atmosphere in about a week. But our warming carbon dioxide gas would linger for decades, or even centuries.

Scientists are getting closer to understanding some other industrial dusts, too. One climate team recently analyzed a variety of natural and industrial particles found crossing the Indian Ocean. The team concluded that all the breeds combined chilled the surface of the ocean below by 16 watts per square meter—a considerable bite out of the global average of 240 watts per square meter. Most sobering, the team determined that of all the airborne particles, the human-made stuff—sulfur, soot, ash, nitrogen compounds, and organic particles—provided twice as much cooling as the natural ones.

And these dusts warmed the atmosphere where they reside by nearly the same amount. Soot in the air proved particularly influential. Although it accounted for a small fraction of the dust bits, this particular breed of dust did more than half the atmospheric heating. And at least for one type of tropical cloud, soot is proving fatal: in the daytime the layer of black dust in the air over the Indian Ocean heats up enough to alter the flow of moisture through the atmosphere. Starved of ocean water, the local "trade cumulus" clouds evaporate, and the Sun bears down unhindered. Scientists have dubbed this phenomenon "cloud burning."

From a cloud's perspective the modern veil of human-made dusts is making *every* day an extra-dusty day. And now there is photographic evidence to prove that our dusts are changing the weather.

Daniel Rosenfeld is the researcher who used a NASA satellite to show that smoky clouds over Kalimantan, Indonesia, fail to rain. Now the Israeli scientist has turned his attention to those clouds that encounter strictly human-made dusts. He has developed a method for identifying polluted clouds in images taken by the satellite.

Rosenfeld first observed a similarity between the smoked clouds of Indonesia and clouds that had passed through the polluted air over Manila, the Philippines: they both seemed unable to produce rain. Next he focused more closely on individual sources of pollution and on the clouds that form near them. What he discovered he calls "pollution tracks." His images show wind-curved plumes of cloud spreading downwind of industrial sites.

Like bright ship tracks and contrails, pollution tracks are clouds that form on a stream of polluting dusts. And Rosenfeld's images trace them back to their source with damning clarity: one track starts at a metal smelter, another begins near a brown-coal power plant, a third strikes out from a cement plant and a fourth from an oil refinery. In Turkey, in Canada, in Australia—Rosenfeld found these great stripes of pollution-born clouds everywhere. Well, except in the United States. He couldn't find pollution tracks over the big cities that cluster in the northeastern United States, he explained, because the air is so thoroughly polluted that the tracks merge into one big curdled mass.

Rosenfeld now wondered if these pollution tracks might also be changing the patterns of rain- and snowfall we've come to rely on. He dug into regional rainfall records. And there he found bad news: compared to their surroundings, the areas under his pollution tracks are rain-deprived.

Rosenfeld's work came as an unwelcome confirmation of another observation. A couple of years earlier, scientists had determined that weather in the United States had begun to reflect the workweek. All of our workday labors—

our driving and combusting and manufacturing—build up to a dust climax at the end of the week, the study concluded. Then, superbly stocked with sulfur beads and other dusts, the skies tend to drop a little extra rain on Saturdays. The dust settles during the nonproductive weekend. So it shouldn't be surprising that statistically, Monday is the least rainy day of the week. The dust of our own industriousness is indeed making its mark on the weather.

The struggle to comprehend how our modern-day dusts might shift the climate makes one nostalgic for the bliss of more innocent—or ignorant—days. In his Seattle office one day Steve Warren dug a vintage research paper from his file cabinet and proffered it with a grin. The scholarly article from the blithe 1970s featured a drawing of a jet spewing four wide trails of black powder into the air. The authors, excited about the heat-absorbing potential of dark-colored dust, proposed that loading the sky with carbon dust would help raise the global temperature to a range more enjoyable for humanity and for crops, too. The wonderful fringe benefits, they enthused, could include enhanced precipitation, interference with cyclones, cooler days and warmer nights, and earlier springtime snowmelt. There seemed no limit to the wonderful things we could do with dust.

8

A STEADY *DOWNWARD* RAIN OF DUST

What goes up must come down. In keeping with that rule, the billions of tons of dust that blow up into the sky each year must eventually settle. And this is bound to alter the face of the planet as the years go by. Falling dust builds richer soils and feeds tiny life-forms on both land and sea. But some settling dusts deliver disease and death. They're killing coral in the Caribbean. They're poisoning food chains. They're settling directly into our bodies.

The dust catchers are prepared for all possibilities. With dust traps on land and deep in the sea, with ice drills and chemistry kits, they are rounding up dust. And they are starting to sort the good from the bad.

Dust catcher Daniel Muhs, a USGS soil scientist, had a hunch about a mismatch between some Caribbean islands and their soils. For decades scientists have been looking for a way to explain why some islands have gray bedrock and vibrant red soil. As far-fetched as it sounds, Muhs suspected that the soil had fallen out of the sky.

Many of the Caribbean islands, Muhs explains, are made from the fossil skeletons of old coral reefs. Consider Barbados, a glorious green paradise in the southern Caribbean. Perhaps a million years ago corals first colonized this shallow spot. They set about building their calcium carbonate housing. The corals spread over a vast field, generation building atop generation. What the corals didn't foresee was that their entire shallow spot was uplifting, pushing them out of the ocean at a rate of about an inch and three-quarters every century. As the reefs breached the sea's surface, a white island was born. Over hundreds of thousands of years this cycle repeated: coral built a new limestone fortress around the nascent island, only to be shoved out of the water. The white island grew. But few plants would have grown there.

"A reef is pure calcium carbonate," Muhs says. "Calcium, carbon, oxygen. Not the right ingredients to make soil."

Nonetheless, European explorers reported that the unpopulated island was handsomely forested when they claimed it in 1627. About thirty years later Barbados was lush with sugarcane. And its current crop, lavish flower gardens, appear equally well fed.

The soil of Barbados ranges from sixteen inches to a yard in depth, and it is orange and brown. It sits atop the old pale gray coral. The soil is mostly "aluminosilicates," minerals rich in aluminum and silicon. The coral is nearly pure limestone, or calcite. And, as Muhs observes dryly, "You can't get aluminum out of calcite."

Previous attempts to explain this condition hewed to the belief that a soil must be born of the rocks upon which it sits: The Barbados soil, therefore, must be comprised of minerals incidentally trapped in the original coral as it grew. As the coral had been shoved out of the water, this theory went, the calcite had weathered quickly away. The grains of dark minerals trapped in the old coral were freed. They accumulated atop the old coral and eventually formed a soil.

Muhs considered this. Then he calculated that this explanation would have required that perhaps seventy feet of reef be eroded away to produce the amount of dark soil that remains. And the fossil reefs looked pretty uneroded.

There was another popular explanation for the mismatch: the Caribbean is littered with volcanoes like the one that ashed over Montserrat in a previous chapter. In fact, in 1979 a volcano on nearby St. Vincent tossed ash a hundred miles to Barbados.

Muhs, however, suspected a different donor. He had read about the mid-1960s attempt to catch space dust in Barbados—the experiment that had caught the Sahara instead.

Scientists already knew that the Sahara threw its dust great distances. As early as 1845 the wandering naturalist Charles Darwin had published a list of dustfall reports by his contemporaries. Near Africa some sailors reported dustfalls so thick that visibility was nil and boats ran aground. And additional reports of falling dust that stuck to sails, decks, and instruments came from far out into the Atlantic, including one from a ship that was halfway to South America when it was dusted. Darwin also reported that microscopic analysis of the dust had produced dozens of species of freshwater diatoms and a similar number of phytolith types.

But how to prove that Saharan dust could add up to a thick soil many thousands of miles downwind? The task was complicated by the tropical conditions: high rainfall quickly dissolves some of the elements in desert dust and carries them down into the underlying stone. This alters the chemical "fingerprint" of the dust. Comparing this weathered dust with fresh Saharan dust would be like comparing photographs of an old man and a boy: would the passage of time have left enough details unchanged that Muhs could be sure the two faces belonged to the same person?

In fact, it had: Working with some of the least soluble elements in the soil, Muhs was still able to tease a distinct fingerprint from samples of the Barbados dirt—it was the mark of the Sahara. Furthermore, based on the age of the coral under the soil, scientists have been able to estimate that the Saharan dust has been tumbling down on Barbados for at least three-quarters of a million years.

Although Barbados is among the southernmost of the Caribbean islands, Muhs has found some indication that the Sahara has long been dusting the Florida Keys, too. And he tested Jamaica, the mid-Caribbean island famous for its music, its marijuana—and its aluminum. Many of the soda cans in the United States are made from Jamaican aluminum ore, or bauxite.

"Bauxite is a very aluminum-rich clay," says Muhs. "It's very weathered: all the other elements have been leached away by rain. Because aluminum is not a very soluble element, it tends to stick around, even under high rainfall. Jamaica has some of the most important bauxite deposits in the world." But again, the rocks underlying those aluminum-rich clays are made of old coral. "To make bauxite you have to start with rock that has aluminum in it," Muhs repeats, "and those limestones just don't have it."

Saharan dust, however, contains plenty of aluminum. "My guess is that the bauxite on Jamaica and Haiti are still-older dust deposits that have been weathering even longer than the soils on Barbados," says Muhs.

Consider this the next time you pop the top off a beverage can. Hundreds of thousands or even millions of years ago the aluminum that makes up that container flew, one speck at a time, out of the Sahara and across the Atlantic Ocean to the Caribbean.

"It's incredible," Muhs marvels. "It really is." Raise a beverage in recognition of the restless Sahara—and then put something over the hole to keep the dust out, because it's still coming.

Dust catchers have found this beneficent Saharan dust settling as far west as mainland South America. Though the fingerprinting of South American soils is just getting under way, most dust scholars already feel confident that Saharan dust is very important to the health of the Amazon basin. A similar analysis of Hawaiian soils shows that Asian desert dust may account for 10 or 20 percent of the uppermost layers. And this dusty rain is a good thing, says Muhs.

"Precipitation is so high in rain forests that the nutrients get leached out very quickly. And of the nutrients that are left, a lot are tied up in the plants themselves," Muhs explains. "These forests just don't have a lot of goodies for plants to use. So how can you sustain a forest for millions of years? It's quite possible that Saharan dust supplies a large component of the nutrients in northern South American soils."

Muhs isn't the first person to notice that dust acts as a gentle fertilizer. About eight thousand years ago Chinese farmers stumbled on this fact. They discovered that there are no troublesome rocks in loess, those deep patches of settled desert dust. They found that loess is child's play to till. That it soaks up water like a sponge. That it grows crops like crazy. That they could dig their homes into the loess hillsides. Today the Chinese loess is terraced into narrow fields that wrap like corduroy around the hills.

Farmers the world over have since made similar use of the loess deposits in their backyards—some of which are still growing deeper. Downwind from the Taklamakan and the Gobi, parts of northern China receive about fifty tons of dust per square mile each year. The dustier parts of the United States—the Southwest and the Great Plains—may receive anything from five to forty tons per square mile. In fact, if you map the "cereal bowls" of the big continents, you will simultaneously map those patches of land that catch the most settling dust. The midwestern United States, the Argentine Pampa, much of Europe, the Ukraine, Central Asia, and a full one-fifth of China—altogether the dust soils of the world grow one-fifth of the planet's wheat, as well as the grains and grasses that feed livestock.

So settling dust doesn't feed just the palms and lilies of Barbados. It feeds us all.

Even in the McMurdo Dry Valleys of Antarctica, one of the most lifeless-looking places on Earth, downfalling dust can inspire a modest bloom of life. Those freeze-dried deserts are home to a handful of sizable lakes. Left over from a warmer time when the ocean advanced on Antarctica, the lakes never feel the wind rippling their surfaces. They're entombed under ten or twenty feet of ice. And in the middle of that ice can be found tiny, dusty bubbles, inside of which little worlds turn in quiet isolation.

"I had been working on the water of one lake, Lake Bonney, looking at plankton and gases," says dust catcher John Priscu, a biologist at Montana State University. "When I started extracting gases from the ice covering the lake, I found that nitrous oxide was concentrated in the same area as a layer of sediments. I looked more closely and found these microbes." These bacteria are the minuscule citizens of sand-size worlds suspended in a lake-size universe.

When dark-colored dust falls out of the sky and onto the ice of Lake Bonney, it absorbs sunlight. It melts a little hole in the ice and sinks. The next day it absorbs a little more sunlight and drills a bit deeper into the ice. And so it goes, until the dust reaches a point about halfway down to the entombed lake. There, under six feet of ice, is a layer of scattered dust clumps, the biggest of which are a few inches long.

One September, before the austral-summer sunlight warmed the ice, Priscu and colleagues cut a trench down through Bonney's ice lid. Pointing a camera into the ice, Priscu caught the frozen universe of the dust layer. In the photograph, strings and clots of dust and sand hang in the blue ice, surrounded by frozen curtains of silver bubbles.

When Priscu put samples of the dusty ice under his microscope, a series of individual dust planets appeared, linked delicately together by bacterial filaments. And in a still-closer shot the inhabitants themselves appear. On a single gray world about as wide as a hair, chemical stains have revealed bright spots. The indigo stain that coats much of the dust grain marks live bacteria. A few streaks of magenta stain crossing the grain indicate concentrated chlorophyll. Enlarged one more time, the magenta streaks finally resolve into neat chains of life. In all, Priscu found about twenty strains of bacteria and ten strains of cyanobacteria, or blue-green algae.

Later that austral summer, as the sunlight strengthened, the frozen worlds over Lake Bonney came to life. Even under six feet of ice, the dark dust absorbed sunlight. Slowly, each clot warmed the ice around it, creating a teensy bubble of water. Given water, the bacteria could awaken from their hibernation and continue the work of breaking down dust and building up generations. "They're just like oases in a desert," Priscu says, "these little pockets of liquid water that support life." For about four months, while the Antarctic sunlight bore down on them, these ice dwellers thrived. Then the Earth tilted them into darkness, and these tiny dust planets froze once again.

The wind drops such a variety of dusts onto the ice that there's bound to be something to suit the taste of even the most discriminating bacterium. A partial census of organic detritus in one short section of Greenland ice, for instance, held traces of fifty-seven different organisms, from bits of trees to fungi, algae, and fragments of other simple life forms. Another count, in a different section of ice, found evidence of two hundred species of fungi alone, along with an identifiable chunk of a common plant virus.

If it seems incomprehensible that a bacterium could adapt to living on dust under six feet of ice, the Lake Bonney bugs are sissies compared to the other ice bacteria that clamor for Priscu's attention. Lake Vostok is a Lake Ontario–size water body on the Australian side of Antarctica. It's buried under two and a half *miles* of creeping glacial ice. In the 1990s an international team slowly bored an ice core down toward the lake. Climate change was the primary interest of these investigators. But the sections of ice were also routinely tested for bacteria. And just four hundred feet above the liquid lake, the team hit an explosion of life.

"Well, I have seven different types of bugs," Priscu says modestly. A colleague has more of the bugs and has cajoled them into growing in his laboratory. This lively ice was part of a seven-hundred-foot-thick layer of refrozen Lake Vostok water that underlies the glacier. And if the first samples are any indication of how densely bacteria populate the ice, it's pretty busy down there. Each liter of melted ice might hold about a million bacterial cells. Compared to seawater this is sparsely settled territory, but given its location, the bacteria's reproduction rate is noteworthy.

Vostok has been sealed over with ice for about 15 million years, afterall, as a glacier has crawled over its head. The pressure of the ice alone, equivalent to three tons per square inch, would seem discouraging. So how did these bugs get here, and what are they eating?

"I think they landed in snowflakes, a half million years ago," Priscu says, betraying no reservations about the durability of his subjects. The bacteria, he proposes, became incorporated in a glacier. These particular bugs happened to land in ice that was flowing toward the buried Lake Vostok.

"They worked their way down and coastward," Priscu says. "When they hit water a half million years later, they revived. People have revived million-year-old bacteria from amber and from permafrost. It's not outlandish. After all, freezing is the best way to preserve bacteria."

Once they revived, Priscu thinks the bacteria may have feasted on dust that also made the long journey from the surface. As Lake Vostok's water circulates slowly, some of its water freezes onto the icy lid. But Vostok's circulating water *melts* another part of the glacial lid, Priscu theorizes, releasing a constant trickle of long-held bacteria—and the dust that feeds them—into the lake.

Unfortunately, we will have to wait six or eight more years to learn the secret to these dust eaters' success. The drilling at Vostok has been halted temporarily to prevent contaminating the lake with mundane bacteria from the Earth's surface.

Like the ice-enclosed worlds that John Priscu found in Antarctica, a glob of "marine snow" is a dust-rich oasis in an otherwise bleak environment. Mary Silver, a marine scientist at the University of California at Santa Cruz, is a dust catcher who sieves the oceans for settling collections of dust and detritus.

These oceangoing dust bunnies often begin with a tadpole-size beast called a larvacean. This creature spins itself a diaphanous "house" of mucus that's intended as a food screen: particles too big for the larvacean to eat will stick in the mucus, while smaller foodstuffs will flow through. But over the course of a month or so, the bigger dusts clog the screen. "And when particulate plugs all the holes," says Silver, "he abandons the house." Perhaps the size of a basketball, the house is murky with everything from desert dust to the manure of tiny crustaceans called copepods to bits of dead zooplankton and diatoms. It is, as Silver puts it, "loaded with yummies."

As it sinks unhurriedly through the dark water, this house may join up with other drifting clots of slime and detritus. A single blob of marine snow will occasionally grow to beanbag-chair dimensions. Even at that size, running your hand through the blob would suffice to demolish it, sending fragments quaking into the surrounding ocean.

Silver looks at the smallest components of marine snow and finds that each particle is a small world. "The particles get organic coatings, and then everything else sticks to them very quickly," she says. "It's exactly like a dust ball, in that everything sticks to it."

Many of the things that stick do so on purpose. A variety of algae colonize the sinking dust specks. Protozoans, slightly more complex organisms, also move in. More sophisticated still, multicelled entities known as metazoans and sarcodines represent the apex of life on sinking dust.

Silver intercepts the falling dust bunnies with traps anchored as deep as a mile and a third in the ocean. And she has found that microorganisms seem to use the sinking dust as elevators, going only as deep as suits their lifestyle. While some creatures can function fine at extreme depths and in total darkness, others require sunlight and gentler pressures.

"The species on the particles change as you go down, so they either die or hop off," Silver says. "If they don't, they're hopelessly lost." Or hopelessly eaten. Even at more than a mile down, a speck of dust can be a very lively place. Protozoa taken from these deep dust specks had "stomachs" filled with bacteria and smaller protozoa.

Marine snow is one "elevator" that can carry dust down to the sea floor. But even dust that fails to entangle in the mucus of a larvacean house has good odds of sinking, although these specks are generally too small to plummet unaided. They may find themselves stuffed into the maws of some small, omnivorous creature. Specks that prove nonnutritious, such as desert dust, will simply pass through, to emerge packaged in a fecal pellet. Though these pellets may be just a few hair-widths in diameter, they're big enough to sink, and they rain down through the water in enormous numbers—hundreds per square foot per day. This is the exactly the unglamorous end that space-dust guru Don Brownlee bemoaned for the precious sprinkling of cosmic dust that falls into the ocean. Mary Silver laughs.

"That's real life," she says.

Dust catchers have so vividly demonstrated the fertilizing effect of settling dust that some farmers have reached the curious conclusion that they're not catching their fair share. To make up for the shortage, they go hunting for dust they can bring home and scatter for themselves. David Miller is a plant physiologist and biochemist at Oberlin College in Ohio who has taken a level-headed look at "remineralization," the practice of spreading dust on farms and gardens.

The remineralization craze sprouted in Germany and is still best known in Europe. The dust flew furiously in the 1970s, when one John Hamaker championed it in a book called *The Survival of Civilization*. "I've never met the guy, but I'd characterize him as something of a zealot," Miller confesses. And like any good zealot, Hamaker attracted followers. All over the Internet, remineralizers now make claims for monstrous vegetables jammed with healthful minerals and trees that shoot up at dizzying speeds, once dust is stirred into the soil. One report even describes the fatter lambs and sunnier egg yolks that resulted when farm animals were given access to dusted grass. These testimonials thrill the rock-crushing industry, which produces between 100 and 200 million tons of useless dust each year.

"There are all these quarries that have all this dust stockpiled," Miller says. "The U.S. Department of Agriculture began calling it 'mineral fines,' to make it more sexy." The USDA even held a conference on remineralization in 1994, trying to stir up interest. Miller and other researchers have planted various test plots, some dusted and some not, for comparison.

"To tell you the truth, I have no doubt it will be useful. But proving it—that will be the hard part," Miller says. "I've got these plots that are ten and eleven years old, and I still haven't seen a change in the mineral content of the plants."

But he also says that other scientists' experiments have shown that dust can provide a quick shot in the arm to seriously damaged soils. He suspects it may be just what the doctor ordered to neutralize soils soured by too much acid. Powdered limestone, after all, is already used universally by farmers for just that purpose. And in extremely nutrient-poor soils an application of fresh dust can change the mineral picture rather quickly. But how that might translate into bigger, healthier plants isn't exactly clear.

More recently, Miller has been looking at how dust functions when it's added to compost. He thinks it may speed the bacteria in their work of breaking down dead plant tissue, and that the bugs may break the dust into optimally tiny bites of plant food.

The USDA, too, has tried adding dust to compost, says federal soil researcher Ron Korcak. But the results were inconsistent, and the notion has simply failed to capture the imagination of the broad gardening public. "Now, drywall . . ." ventures Korcak, who has experimented with the dusts of many industries. "I think there are a lot of benefits to taking scrap drywall at a construction site, pulverizing it, and adding it to the landscaping." This hasn't caught on, either.

Nevertheless, the die-hard fans of dust aren't waiting for government-approved, peer-reviewed research papers to tell them how to grow supercarrots. They're digging the dust into their farms and gardens. Glacial dust is the choice of purists, since it contains a sampling of all sorts of rocks. Granite dust is also admired for its diversity of minerals. But in terms of availability, it's hard to beat whatever's being crushed at the local rock quarry. For such dust, remineralizers pay a few bucks a ton and apply it to fields at a rate of a few tons per acre.

When dust settles, it doesn't always deliver such blessings. Some dust catchers are intercepting specks that rain down disease and death.

Geologist Gene Shinn has been photographing the coral reefs of the Florida Keys since the 1950s. Today something is clearly amiss. The sponges are there, knobby vases of gold and purple and orange. Schools of bright fish dawdle over the reefs like giant butterflies. Conchs shuffle ponderously through the white sand of crumbled coral. But the brain corals, those gently tinted boulders? Many are ringed with black-band disease. And purple sea fans, lacy sheets of polyps stretching five feet off the reef, carry the bull's-eye marking of a deadly fungus.

Shinn, who works for the USGS, says he began to notice signs of coral disease on the reefs in the late 1970s. "Then, in the summer of '83, two major species of branching coral that had begun to suffer in the 1970s died—ninety percent of them—all over the Caribbean," Shinn says in his gravelly voice. "And a species of black sea urchin that kept everything clean died within just three or four months." And lavender-colored sea fans developed dark purple ulcers that spread outward, leaving dead tissue in the center.

Granted, nearly every coral reef in the world is suffering today. Diseases and bleaching are pandemic, and algae are taking over the hardest-hit reefs. Theories abound, the most popular being that a combination of pollution and warming oceans is simultaneously weakening corals and strengthening their enemies. So when the corals of the Florida Keys began to suffer, Shinn and most other coral researchers assumed they were suffering from human pollution. Human sewage, animal waste, and crop fertilizers that are generated on land near the coast all tend to find a way to the sea. And these so-called nutrients can feed a bloom of algae that overrun coral.

Pollutants and climate change probably *do* play a role in the war on coral. The die-off of algae-eating sea urchins compounds the problem in the Caribbean. But Shinn, in the wake of the coral epidemics of the 1980s, became intrigued by the work of two dust scientists. In one ear he heard Miami dust tracker Joe Prospero saying that the Sahel drought started kicking extra Saharan dust across the Atlantic in the 1970s and that the falling dust had become a downpour by 1983. Prospero estimated that hundreds of millions of tons were crossing the Atlantic each year. In the other ear Shinn heard one of the iron-fertilization scientists report that dumping iron into the ocean causes algae blooms. "I started noticing that these disease events happen in years with maximum dust flux," Shinn says. "And 1983 was a very dusty year."

Initially, Shinn proposed that the Saharan dust was turbocharging algae by feeding it iron and other nutrients. But now Shinn's theory has changed. It has since grown into a major investigation of all the infectious microbes that might be accompanying Saharan dust to the Americas.

In 1996 a team of South Carolina biologists identified the DNA of a fungus called *Aspergillus sydowii* in the sickened sea fans. And they found the same fungus in Saharan dust caught flying over the Virgin Islands. They intentionally infected isolated sea fans with the Saharan fungus, and the disease bloomed before their eyes.

Various *Aspergillus* fungi live in soils all over the world. Tough enough to withstand the preservative effects of both salt and sugar, *Aspergillus* is known to most people as the spots on top of old jelly. But while it's ubiquitous, it's not completely harmless. One species of *Aspergillus* can cause deadly lung infections in the elderly, people with HIV, and others with weak immune systems.

It may seem silly to suggest that *A. sydowii* could leave a happy life in the parched African Sahel and take up residence underwater in the Caribbean. But to microbiologists, such adaptability is unremarkable. Once the fungus is snared by a sea fan, it does seem to thrive. Some fans recover from the infection, but many die. Because the fungus hasn't mastered underwater spore production, Shinn says, an *A. sydowii* infection cannot hop from one sea fan to the next. So to keep the epidemic going, fresh spores must be supplied frequently to the Caribbean.

Some scientists take issue with Shinn's dust-delivery method, proposing that *A. sydowii* could as easily flow off eroded land in the United States, to be distributed around the Caribbean by water currents instead of by the wind. Shinn says the pattern of the plague argues against that.

"When these events happen, like the death of the sea urchins, they happen almost simultaneously, everywhere in the Caribbean," he says. They don't leapfrog from one reef to the next, the way you might expect disease to spread on ocean currents. "If it all fell from the air, in one big cloud, then it makes sense," Shinn concludes.

Furthermore, sea fans are struggling against the fungus even around some of the least-peopled Caribbean islands—islands that can offer neither erosion nor pollution. "It's around isolated islands where there's *nothing*," he says. "Navassa, west of Haiti, has visibility of two hundred feet—the water's crystal clear. Not a soul lives there. And the sea fans are sick."

To make matters worse, the iron and phosphorous that flavor Saharan dust probably *are* a boon to algae. And if algae spread faster than snails and fish and sea urchins can eat them, they can march right across a coral, essentially smothering it. Algae can also monopolize the coveted hard surfaces of a reef, leaving new corals no place to take root.

And now Shinn wonders if Saharan dust might be attacking people as well. People in the Caribbean sometimes report sinus headaches when Saharan dust throws a haze across the water and settles pinkly on their boats and cars. Even in South Florida, doctors are increasingly aware that an incoming cloud of Saharan dust may hit hard in the lungs of people with asthma and other breathing difficulties. Shinn has cautiously noted that the beginning of the recent Saharan dust deluge coincides with a skyrocketing asthma rate, which also took wing in the 1970s.

"I'm a geologist. I just stumbled into all this," Shinn says with a self-deprecating laugh. "But when I go out and give talks about this, it gets people on the edge of their chairs. It has horrendous potential. I've been in Puerto Rico when they've had one of these dust events," he continues. "You can look right at the Sun, because it's so hazy. You can smell it. It smells like dirt—like a dusty road. You can *feel* it. They close the airports. We already know Saharan dust is loaded with *Aspergillus*. But the truth is, no one has ever tried to culture the other stuff that must be in it. We want to see what else is in there."

Since he began investigating Saharan dust, Shinn says he's come across a number of strange tales: a cistern on a Caribbean island fills quickly with dust, and that dust is unexpectedly rich in toxic mercury—which, incidentally, is mined in the Saharan nation of Algeria. Another cistern yields dust fraught with pesticides long banned in the United States but still used in other countries. In 1989 the windward islands of the Caribbean got pelted with millions of inch-long locusts, along with the African dust. "I'm just full of anecdotal stories like that," the dust catcher says, laughing. What he needs, though, is hard data.

In the spring of 2000 NASA decided to help him get it. On February 26 a NASA satellite photographed the most monstrous cloud of gold African dust that anyone could remember seeing. The dust mushroomed westward, over the Atlantic. In the Azores a scientist captured samples and sent them on to Shinn and colleagues for analysis. The levels of mercury and radioactive beryllium 7 stunned the researchers.

"We couldn't *believe* how high the beryllium seven was," Shinn says. "One sample was three times the upper limit for the workplace. Mercury was two parts per million—it's normally forty or fifty parts per *billion*. They're breathing this stuff in the Caribbean! The radioactive particles get in their lungs!"

The beryllium 7 is produced naturally in the atmosphere. And normally, it stays up there. But flying desert dust spends so long in the sky that it probably soaks up the radioactive beryllium, then brings it down to Earth. The source of the mercury isn't known. This dust cloud was also high in radioactive lead, a natural product of radon.

The space agency decided to sponsor a major dust-catching campaign in the Caribbean. NASA wants to know what the dust does to coral reefs. But the agency also wants to know what dust might be doing to the millions of people who live in the islands, and in the southeastern United States. At last dust catchers will be able to make a census of the life forms and chemicals that plummet Earthward with Saharan dust.

One microbiologist has already begun to take a head count. Garriet Smith is the University of South Carolina researcher who identified *A. sydowii* as the sea-fan foe. And now, with federal health funding, the marine biologist is scanning Caribbean-caught Saharan dust for additional killers.

"They want us to do the fungi first," says Smith, a gentle-voiced man with a long silver ponytail. "because they potentially represent a public-health threat." Fungi are a threat to people because they're immune to most drugs. A healthy body usually fights off fungi successfully. But if these invading dusts get a foothold, they're very difficult to dislodge. For people with weakened immune systems, fungal infections are often fatal.

And there are fungi aplenty in the Saharan dust. When Smith isolates their spores from the dust and drops them on a plate of fungus food, they spring to life. A very general DNA survey has turned up a broad diversity of soil fungi, including a blastomycete, which can produce a pneumonialike disease.

Bacteria in the dust will be Smith's second priority. "There's quite a variety of those as well," Smith says, though he hasn't had a chance to identify the individuals. The whole notion of contagions plummeting from the sky is adequately novel that few people are working on it yet.

"It's fun, whenever we find something," Smith confesses with a soft laugh. "Because everything's *new*. Nobody's ever *looked* before." As NASA's investigation picks up speed, however, more people are looking. In the spring of 2001 Gene Shinn reported that his team had found 139 bacteria and fungi in the dust, including 14 species that attack crops such as wheat, beans, elm grass, and peach trees.

To some dust catchers the threat of falling microorganisms is new and alarming. To farmers it is *old* and alarming. Unlike desert dust, mold dusts intentionally take to the wind, with the goal of settling down on a fresh victim. And the successful settlers are responsible for nearly three-quarters of the damage done to major crops in the United States, costing farmers billions of dollars a year.

"Blue mold" is one such dust. Tobacco farmers often spray once a week to fend off this menace, making tobacco one of the most chemical-intensive crops on the planet. Charles Main aims to take some of the mystery out of blue mold and reduce the need for spraying, by predicting where and when this dust will land. A semiretired plant pathologist at North Carolina State University, in the heart of tobacco country, Main has been tracking the rain of fungal spores for so long that he's now able to forecast it like weather.

"We're pulling together meteorology and biology," he says. "We can give people forty-eight hours' warning now."

Main works with a network of mold spotters in Cuba, Central America, Canada, and the United States to identify blue-mold outbreaks. Then, using weather-forecasting computers, he can predict where blue mold will next rain down. One spring day Main said he had just spotted the year's first incursion of mold onto U.S. soil. As usual, the fungus had dined on Cuban tobacco and then hopped north. "Friday we saw winds out of Cuba taking the mold toward Mexico," Main said, "But Saturday the winds turned and came over South Florida." From Florida, if the spring winds followed their normal course, the mold would leapfrog north, clear to Canada. And Main would stay one step ahead, basing his forecast on both weather and the mold's vulnerability to that weather.

"These particular spores are very ephemeral," Main says. "In direct sunlight they probably wouldn't last thirty minutes. When the spores start out from Cuba, if there are thunderstorms, they'll get washed out of the air. If there's bright sunlight, most of them will be dead when they get to Florida. But if it's cloudy or foggy, they'll survive. So we look at the weather all along the projected track."

Because the hazards of air travel claim so many spores, all molds must produce daunting numbers of them. Some infested fields can fling a billion mold spores into each cubic yard of the surrounding air. Sky rivers of fungi are constantly crisscrossing the planet, often on predictable schedules. Many of the travelers will die in flight. And many will fall in pointless places, miles from their favorite plants. But many more will land in the perfect position to keep the plague moving.

One of the most Viking-esque mold clans is the "rusts." Attacking everything from cereals to apple and coffee trees, the rust fungi cause some of the world's worst plant diseases. And they're great travelers. In 1966 a coffee rust ravaged Angola, on the Atlantic coast of Africa. Then it hopped the Atlantic in about a week, to savage the coffee trees of Brazil. Wheat rusts think nothing of departing Mexico to settle on fields many hundreds of miles away in the northern United States.

Farmers' eagerness for a forecast that can warn them of an approaching threat is evident at Main's Web site, where he updates his blue-mold predictions three times a week. The site gets at least a quarter million hits a year, as tobacco farmers throughout Central and North America check the dust forecast before they deploy the pesticides.

And Main isn't satisfied with forecasting blue mold. He works with Estelle Levetin, the Ohio aerobiologist, to produce pollen forecasts. He anticipates that predicting the paths of these floating allergens will be just as successful as forecasting the tobacco plague. "By the time allergens are caught and counted," he says, "patients are already sneezing. We think we can provide enough lead time to take protective action." Dust forecasting is becoming a reality.

. . .

Forecasting is one way to get a better grip on settling dust. But "hindcasting" can be useful, too. Some dust catchers are using settling dust as an indication of which dangerous dusts are *rising* somewhere else. Steven Eisenreich, a Rutgers University environmental chemist, oversees a network of dust detectors that sieve the air of New Jersey. From pollutants they catch, he can tell you when the wind is from the south: the termite pesticide chlordane blows into New Jersey. And when the wind is from the north, the air is clean.

Many of the "persistent organic pollutants," or POPs, that Eisenreich snares travel as gases. These gases may fall in rain, or they may condense onto other particles in the air, but generally, they get around in gas form.

Consider polychlorinated biphenyls, or PCBs. This family of toxic chemicals builds up in animal fat, gradually making its way up the food chain. Most members of the PCB clan enter the waters of the Hudson River estuary as gas, sliding across the line between air and water one complicated molecule at a time, says Eisenreich. But some of the bigger molecules in the PCB family may travel through the air en masse, as particles. And like all flying particles, they must eventually return to Earth. PCBs, many of which leaked out of electrical equipment decades ago, endure so long in the environment that the same old PCBs are still playing musical chairs: A particle is lifted off the ground, it blows around, and it settles somewhere else. A day later, or a year later, it takes flight again and settles somewhere new.

"Where are all the PCBs we've released over the last fifty-five years?" Eisenreich asks. "They're in the soils and vegetation of North America, Europe, the Arctic, and the waters of the Atlantic. What we don't know is how long the soils will keep on giving up those PCBs to the air." Eisenreich actually *wants* the PCBs to fly into the air, because the atmosphere seems to be the best place for natural chemistry to break them into smaller, safer molecules. But only a few molecules are destroyed on each flight. And only a few hundredths of a percent of the world's PCBs are in the atmosphere at any given time.

PCBs are just one family of the industrial chemicals circulating between Earth and sky. The United States banned leaded gasoline decades ago, but lead is still stored in roadside soils. It is free to go for a ramble whenever the wind invites it. Mercury and dioxins, arsenic and DDT also circulate tirelessly and rain down continuously. Whether these toxic substances take a sabbatical in the soil when they settle, whether they lurch back into the air, or whether they're sidetracked into the food chain—that's the luck of the draw.

The best picture of how thickly toxic dusts rain onto the face of the Earth comes from the first monitoring network that Eisenreich built, which has been catching dust around the Great Lakes for more than ten years. That network showed that about four hundred pounds of PCBs, in the form of rain and dry particles, fell into Lake Huron in 1994. Eighty-one pounds of DDT fell in, too. And more than a hundred tons of mercury, most of it gaseous but a fraction of it in dust form, dropped into the lake in 1994.

Because these toxic chemicals can rise off one nation and then rain down their destruction upon another, they are inspiring international agreements to cut pollution. If one nation hopes to protect its fish and bears, its frogs and its people from the rain of poisonous dusts, that nation must find a way to prevent toxic dusts from rising everywhere on the globe.

Finally, there are dust catchers who attempt to identify the specks that settle neither on land nor sea but in your lungs. At least a billion and a half pieces of dust enter your nose and mouth every day, if you breath exceptionally clean air. Most people inhale many times that number.

The human body is no stranger to dust, of course. Our race grew up in deserts and caves, in pollen-fogged savannas and mold-dank forests. We evolved noses and throats that stop many of the natural dusts before they reach our lungs. So most desert dust and pollen grains never make it past your throat. To an admirable extent, the human body is dustproof.

But in the past century the dust that settles around us has changed. The dusts of the industrial era are often far tinier than the natural specks. They can get past the sticky traps that stand guard in your head and chest. They can penetrate the smallest twigs of your branching lungs and lodge there. Around the world, people are dying of air pollution in numbers that are a bit hard to comprehend—a million people in China each year. Sixty thousand in the United States.

Morton Lippmann has been following the deadly trail of dust for his entire career. He slaps the leg of his brown trousers and looks for the little cloud to rise. "Forty-five years with dust," he says in a moderately muddy New York accent. Although he has a world-weary air, an unabashedly exuberant homemade painting of roses hangs over his desk. Lippmann is an epidemiologist. He analyzes the patterns of disease. He specifically investigates events in which air pollution shoots up and then hospitals and mortuaries *fill* up.

To connect dust and death, an epidemiologist will often record the daily ups and downs of dust in a particular city: perhaps one day there are 40 micrograms of dust per cubic meter of air. The next day there are 50. The day after that 35. At the same time, he records the death rate in the same city, day after day. And when he compares a graph of fluctuating dust levels to a graph of fluctuating death rates, he sees that the two rise and fall together.

"It looks like air pollution kills people," Lippmann says bluntly. "And it looks like particles are the best indicator." One particularly worrisome message from the epidemiologists is that they can't yet detect a level of dust small enough to cause *zero* health effects. Regulators may simply have to accept that wherever they ultimately set the dust limit, some people will still die.

Epidemiologists can get frightfully specific in their predictions of how more dust will bring more death. These studies generally conclude that every time the dust level rises by 10 micrograms per cubic meter—from forty to fifty, perhaps—the death rate rises by a half percent or a percent. "We're talking about small elevations of risk," says Lippmann. "It's small, but cities have millions and millions of people."

Epidemiologists are getting so good at their craft that they can tell you what these people will die of. Lung blockages are common. Deadly respiratory infections rise. And, strangely, congestive heart failure, dysrhythmia, and coronary artery disease also kill more people when the dust rises. In some instances when the dust thickens, these heart diseases appear to kill more people than lung diseases do.

These body counts overlook a much broader haze of illness that dust casts over people. One study has concluded that an extra 10 micrograms of dust in the air causes between 10 and 25 percent more attacks of bronchitis or chronic coughing. Another found that adding 10 micrograms of the dust to the air will inspire 1 to 12 percent more asthma attacks. And a third found that residents of the nation's dirtiest cities may suffer as much as a 15 percent decrease in lung capacity compared to people in cleaner cities.

Hardest hit by dust, Lippmann says, are people over sixty-five and children. Pound for pound, children breathe more air than adults do, so they take in more dust. Their developing bodies are less able to dispense with toxic chemicals in the dust. Epidemiologists have linked extra-dusty air to children's asthma attacks, and even to sudden infant death syndrome. Dusty air also correlates with a lower birth weight for babies, which in turn presages a more troublesome childhood.

"Particulate matter does, in fact, relate to longevity," Lippmann says. "And it seems to be related to SIDS—sudden infant death syndrome. We're pretty sure it's worthwhile investing resources to control particles, even if we don't know all the variables."

And so Lippmann has joined a growing army of dust catchers who are trying to round up scientific support for stricter dust laws. The institution he works for, an unassuming branch of New York University Medical School ninety miles north of Manhattan, is among a handful that received multimillion-dollar grants from the EPA for just this purpose.

When the EPA first established health limits on airborne dust in 1987, the agency's concern was with "PM-10," or particulate matter that is ten microns or less in diameter. These particles, one-tenth as wide as a hair, were thought to be small enough to penetrate into the lung. But soon scientists were arguing that the biggest threat came from even *smaller* particles. A landmark Harvard study estimated that tens of thousands of people in the United States were dying annually from fine dust. And to make matters worse, many parts of the United States couldn't even meet the standard for the *bigger* dusts. In 1999 some 30 million people in big cities and in arid states were still breathing air that exceeded the original PM-10 limit for dustiness.

In 1997 the EPA wrote a new rule that addressed dust smaller than 2.5 microns in diameter—PM-2.5. And in terms of the total amount of tiny dust allowed, the EPA lowered the limit from one five-millionth of an ounce per cubic meter to about half that.

An explosion of protest boiled up out of the industrial Midwest and this howling was joined by trucking organizations, whose diesel-powered vehicles can contribute as much as 30 percent of the tiny dusts that dim urban air. The EPA found itself on the business end of a lawsuit, which accused the agency of regulating without clear proof that dust kills. When a federal court blocked the new dust standard, the EPA went looking for a few good dust catchers and asked them to reel in the proof.

Sitting across the hall from Lippmann is his partner in dust catching, Richard Schlesinger. Schlesinger is tall and terrifically tidy, in pressed wool pants and shiny black shoes. The blinds in his shipshape office are closed. Even his smile is quick and efficient. Schlesinger is a toxicologist—someone who studies the effects of toxic chemicals on living things.

"Particulate matter has been shown—*epidemiologically*—to kill people," says Schlesinger with a wry smile. "But there's really very little, in the way of biological plausibility, to explain why this should be." From behind his mustache comes a small, punctuating exhalation. "The biological explanation is *really* not clear."

By contrast, there are many reasons your body should simply laugh off an attack of dust, he says. Your first line of defense is the hairy, sticky nostril itself. Dust sucked into your nose may get caught there, trapped until you blow into your hankie. If a speck of dust should escape that fate, it will soon find itself whirling through your nasal turbinates, delicate spirals of bone lined with more sticky tissue. Air can negotiate the maze, but the momentum of the biggest dusts causes them to crash into the walls. Beyond the nose, the opening to your larynx presents another obstacle. Dust caught there is carried by a steady flow of mucus down your throat and into your stomach.

But dust smaller than about 5 microns, or one-twentieth of a hair's width, can ride the air past those initial traps. Fortunately, another series of pitfalls lies ahead. Even as air shoots down the straightaway of the trachea, the tiny dust it carries may get stuck on the mucus-coated walls. If not, then each time the bronchial tubes divide and subdivide in your lungs, dust may fail to make the turn and hit a sticky wall. Over the course of a day or two the moving coat of mucus in these tubes will shuttle the captives up to your throat and dump them into your stomach. That is the body's old-fashioned defense system.

It isn't fail-safe, Schlesinger says. When you breathe through your mouth, for instance, air bypasses the filters in your nose and you suck extra dust into your lungs. Furthermore, some dusts have proven capable of slowing the mucous escalator that evicts them. It normally rises at a fraction of an inch per minute, with wide variability among individuals. But certain dusts, tobacco smoke among them, sabotage the mucous machine. This gives trapped dusts more time to react with your respiratory tissues.

This antidust system is also vulnerable to the fine dusts of combustion. The truly tiny dusts can fly straight through the traps, ride the air past the sticky walls, and enter the sanctum of the alveoli, the lung's tiny inflatable air sacks. And in those little balloons, Schlesinger says, some dusts may stay for years. What the dust does in there, that's the $50 million research question the EPA hopes to answer.

"We *could* set the standard based solely on epidemiology," says Schlesinger, whose job is to narrow the roster of suspects. "But it could result in unnecessary expense, in terms of pollution-control strategies. Instead, let's try to figure out *what* in the air causes the effect. If we knew that, say, one specific chemical in dust was the problem, then perhaps we could regulate that chemical, instead of *all* dust."

At the moment toxicologists are thoroughly stumped. There are thousands of compounds in industrial and automotive dust. Maybe all of them are the problem. Maybe two of them are the problem. Maybe the ozone gas that accompanies the dust is *part* of the problem. And that begs the question, *which* problem? Perhaps one dust causes asthma attacks while another causes bronchitis and a third causes heart trouble.

"Cardiovascular disease—heart attack, stroke, stuff like that—how do you get a heart attack from inhaling low levels of particles?" Schlesinger asks with a perplexed tilt of his head. "How do you get that from a particle that lands in the *lung?*"

Researchers are not completely without theories. Some of these address the "flavor" of the dust. Metal-rich dusts, for instance, are a frequent object of suspicion. The act of burning fossil fuels, Schlesinger says, releases chemically reactive metals, such as copper, nickel, iron, and vanadium. Perhaps they react with lung tissue. Acidic particles, including dissolved sulfur and nitrogen beads, are another frequent suspect. They, too, are naturally reactive.

Diesel soot has long been on the list of dangerous dusts. The federal government rates diesel dust as "highly likely" to cause cancer. A California air-pollution agency goes further, estimating that diesel soot accounts for nearly two-thirds of the total cancer risk posed by outdoor air pollutants in the Los Angeles area. Other smokes, of the type that waft from oil refineries, auto-body shops, and automobile exhaust pipes, are strongly suspected of being fatal as well.

But others think the *size* of dust may prove to be the critical factor. After all, the tinier the dust, the more deeply it can dive into the lungs and the longer it can reside there. One gold-standard epidemiological study looked at dust and death in six different U.S. cities. Although the mix of dust flavors was different for each city, the death rate proved nearly blind to those differences. The "dust graph" that best matched the "death graph" was the one that simply recorded the total number of fine particles in the air: as the concentration of tiny dusts rises, so does the death count, seemed to be the simple message.

"All different kinds of chemicals, all different sizes," Schlesinger says, summarizing his challenge. "Coupled with different gases. From anthropogenic and natural sources." Eyebrows raised, he drums his fingers on the desk. Nonetheless, he and other toxicologists are making progress.

Various experimenters have convinced human beings to breathe all sorts of strange dusts for short periods of time. To willing participants they've dished out dusts of iron oxide, manganese oxide, diesel exhaust, and beads of sulfuric acid. To trace the path of air through the lungs they've served up more inert items, including Teflon dust, aluminosilicate dust with radioactive tracers, and microscopic bits of polystyrene.

But animal studies are pointing more clearly toward possible triggers for human death. Exposing animals to specific flavors of dust has produced specific ills. Superfine carbon dust changes blood makeup in a way that raises the risk of stroke, for instance. Diesel soot also alters the makeup of blood, changing both the number of platelets and the white blood cell count. The exhaust from oil burners increases arrhythmia, or irregular heartbeat. Metal dusts cause biological reactions in the lungs that lead to swelling. Exposing animals to "off the street" polluted air has revealed numerous ways that dust can drag down its victims. Taken as a diverse gang, these mixed dusts alter blood chemistry. They change the heart rate. They change the mixture of defensive chemicals in the lungs.

And in both people and animals, scientists have found that a variety of dusts injure the mucous escalator that ships dust out of the lungs. Tobacco smoke is one. But a study of English woodworkers found that they, too, had very slow escalator systems, probably from years of exposure to wood dust. Sulfuric acid beads, one of the most common particles in polluted air, also tamper with the escalator. At first, sulfur actually inspires the mucus to move faster, cleaning the lungs more quickly. But over time the effect reverses, and the sulfur overpowers the means of its own eviction.

These discoveries are suggesting ways in which dust might kill people. Simple irritation and swelling—the standard responses to a bodily insult—may be enough to push weakened lungs beyond their ability to absorb oxygen. The heart-related deaths may be an extension of that failure: a weakened heart may kill itself, trying to pump more oxygen-starved blood to the needy organs.

The dust catchers' quest is growing more urgent. Dust continues to settle inside human bodies, and sixty thousand times each year in the United States, that settling dust brings a life to an end.

The dust catchers are at the beginning of a long process of sorting the falling dusts. Of the countless tons of tiny things that settle around us every year, they are showing us that some fall as fertilizer and others rain down as tiny agents of death and disease. Many of the specks are so new to our awareness that we have no idea what their effect might be.

Perhaps we'll soon realize we should duck and cover when desert dust rains down from Asia or Africa. Not long ago scientists dismissed desert dust and blowing soil as mere nuisances—particles too big and too bland to harm us. But as record-breaking dust storms roll out of Africa, dust catchers in the Caribbean are finding that the dust is extremely fine, and accompanied by some not-so-bland hitchhikers.

As for toxic industrial dusts, we have long known that their descent could cause problems for those beasts and plants and people who lived nearby. But when we watched a murky plume of dust depart the hometown sky, it seemed to disappear harmlessly.

Now the dust catchers are discovering that even as the hometown dust is departing, someone else's dust is filtering in from behind. All day, every day, the dust is coming home to roost.

9

A FEW UNSAVORY CHARACTERS FROM THE NEIGHBORHOOD

Eight women sit in a circle on the floor of a cave in Turkey. On stumpy-legged tables before them, they roll bread dough into wide, paper-thin disks. These they pass to an elderly woman who hunches by a steel pan. The pan perches over a fire hole in the cave floor, into which the old woman scrapes a steady supply of dry grapevines and grass. Each bread disk toasts briefly on the grill, then is laid atop a growing stack. Through the coming winter the circles will feed the hostess—a middle-aged, dark-eyed lady—and her family.

The women talk and laugh as they work. At midday they fill a series of the fresh disks with scrambled egg, cheese, and herbs and fold them shut. They pass these and small, scarred apples around the cave room, which is open on one side to the autumn sunshine and the breeze.

Their faded calico clothes and head scarves are dusted with flour. The thin trail of smoke rising from the fire pit meanders through the cave before departing on the wind. But while both flour and smoke are damaging dusts to inhale, more murderous particles may lie in the thick walls of the cave itself.

Thirty million years ago in central Turkey the local volcanoes threw out an enormous blanket of ash. As the ash settled, it turned to pale pink and tan rock that's soft enough to scrape away with a spoon. Over millions of years wind and water eroded the ash into strange shapes, leaving groups of rock cones that stand fifty or a hundred feet high. From a distance these groves of cones resemble flocks of beige rocket ships, noses pointed skyward. But on closer inspection it appears they have tiny windows carved into them, and inconspicuous doors. Others are split apart by age, revealing the maze of rooms and twisting stairways carved inside by their human inhabitants. For thousands of years people have carved their homes, barns, and churches into this rock.

The cave of the breadmaking hostess stands in the village of Göreme. Although many of the Göreme villagers have abandoned their ancient cones in favor of cement-block houses, some are sticking with their caves. The hostess's broad terrace is carved into the side of a massive cone. Deeper inside the cone the kitchen is dim and cool. Its carved windows boast ruffled curtains, and electric lights glow at the ends of wires running over the rock walls. A box of matches sits in a cubbyhole chipped above an electric stove. Deep underground, the family donkey brays in his carved stable. It is a typical cave home, like countless others scattered across the broad plains of Cappadocia.

Ancient volcanic ash is not normally a big public health concern. The tiny flecks of hot magma usually cool too quickly to form menacing crystals. But just add water. As soon as this ash settled, falling rain began to wear tiny tunnels through it. As the water migrated, it dissolved molecules of ash, then deposited some of those minerals in a new form. Molecule after molecule, the trickling water constructed long, thin crystals that filled empty pockets in the ash. Too small to be seen with the naked eye, these needlelike crystals steadily accumulated, sometimes in single spears, sometimes in fat bundles. And then, in good time, the rain and wind began to slice open the aging ash blanket.

Although this opened the secret of the ash to anyone with the technology to look closely, it wasn't until 1974 that Turkish scientists made their first close examination of the rocks, in a tiny cave village called Karain. They had received word of a strange cancer cluster. Of the eighteen people who passed away in Karain that year, eleven died of mesothelioma of the lung, and three others died of gut mesothelioma. Only four deaths *weren't* from mesothelioma.

Mesothelioma is a very rare cancer. It normally strikes people who have worked intimately with asbestos dust. But there has never been an asbestos industry near Karain. The village is a quiet grove of cones and stone-block houses, nestled into the bare ash-rock hills that edge a flat valley. The people are farmers. Mystified and alarmed, scientists descended. What they found convinced them that the special dust of Karain—and of a few other cave towns—had probably been killing people since they first tapped into the soft rock, countless generations ago.

Karain's very name, Turkish scientists soon reported, means "abdominal pain." The investigators also related an old saying about the village's inhabitants: "The peasant of Karain falls ill with pain in the chest and belly, the shoulder drops, and he dies."

Researchers began fanning out to the other towns in the area. They found a handful of similarly cursed towns, discrete spots of dust-induced misery. One village would report a dismal rate of mesothelioma, while the neighboring village would report none. Karlik, a few miles north of Karain, was cancer-free. Much-touristed Göreme to the northwest also seemed clean. But thirty miles away the larger village of Tuzköy ("salt town") was suffering a mesothelioma rate thousands of times higher than normal.

The hunt for a cause began. In the afflicted towns scientists collected samples of house dust, street dust, whitewash powder, well water, soil, building blocks of the ash rock, and samples of the cave homes themselves. With the help of local doctors, scientists procured samples of lung tissue from people with lung disease. And nearly everywhere the scientists looked, they found microscopic needles of a crystalline mineral called erionite. They found it in lungs, in streets, in the soft rock. A sample chipped from a block of stone in Karain's new library wall proved to be about half erionite, the tiny bundles of rock needles jumbled every which way.

When the investigators' own dust settled, it appeared that at least six villages had the bad luck to have sprung up next to, or dug down into, an ancient wealth of water-forged erionite. Perhaps since the first cave carvers arrived, the people in those villages had been inhaling the malignant dust. The first brown-eyed baby who took a breath in each village was welcomed to the world by that local powder.

More than a quarter century after the discovery of these hot spots of dust disease, with another generation infected by the needles, the Turkish embassy in the United States confessed that none of the afflicted towns had yet been evacuated. Such efforts, the embassy reported, take time.

Wherever you live, the dusts at your front door will have a special, local flavor. And in some parts of the world the hometown dust is a heartbreaker. Some of these malignant dusts are perfectly natural. Germs like the Valley fever fungus and the deadly hantavirus travel as dust. Even humble desert dust can cause lung disease. But to produce local dusts that wreak true havoc, dusts that kill thousands? That's not really nature's realm. That's a job for the industrial revolution.

Our breed was already a smoky little fire starter and a desert-enlarging sheepherder before the advent of engines and powerful machinery. But after the big machines got rolling, every factory town and mining village in the world acquired a distinct dust signature. The industrial revolution gave birth to a cloud of human-made "pollens," industrial dusts, which bloomed again in the lungs of workers and neighbors. Medical slang recorded this flowering of new dust diseases:

Mill fever. Fly-ash lung. Wood-pulp-worker's disease. Aluminum lung. Bakelite pneumoconiosis. Detergent-worker's lung. Meat-wrapper's asthma. Air-conditioner lung. And, of course, asbestosis.

Halfway around the world from Turkey, Libby, Montana, is a crisp little mountain town of about 2,500 people, hemmed in by dark evergreens. As in Cappadocia, people are dying of lung disease at a horrifying rate. But unlike the dust in Cappadocia, Libby's fatal dust was unearthed and set loose by industry.

On a mountain five miles out of town is the broad scar of a vermiculite mine opened in the 1920s. From 1960 to 1990, W. R. Grace & Co., the company made infamous in the book and movie *A Civil Action,* blasted out this vermiculite and took it into Libby for processing and shipment to other factories. The vermiculite, a type of rock that puffs up when heated, was sold for use in cement, gardening products, drywall backing, and insulation.

Health and environmental officials looked in on W. R. Grace's operation over the years, because this particular vermiculite deposit was shot through with thick veins of the crystalline mineral asbestos. Some of the vermiculite contained 40 percent asbestos when it was mined.

Now, asbestos has a long history of *aiding* humankind. Thousands of years before people discovered the risks of handling this dust, they understood its special charms. More than four thousand years ago Finnish people stirred asbestos fibers into their pottery clay and added it to their house chinking. About two thousand years ago Romans wove the fibers into funeral shrouds that would survive the funeral pyre, keeping charred bones separate from the wood ash.

There were early hints that this useful dust was deadly. The Roman naturalist Pliny the Elder observed that the unfortunates who mined and wove asbestos were a sickly lot. He even devised for them some sort of bladder mask. But his warnings didn't reach the other parts of the world where asbestos naturally occurred. About twelve hundred years after Pliny's asbestos alert, the tireless traveler Marco Polo reported finding a remarkable sort of rock in the mountains of what is now western China: he said people there pounded the rock to break it into fibers. These they spun into thread, then wove into an extraordinary cloth. Instead of laundering this fabric, housewives tossed it into the fire, from which it emerged snow-white. Marco Polo and others called asbestos "salamander" or "salamander wool" after the apparently fireproof animal that occasionally crawled out of logs in the fireplace.

It was shipyard workers in World War II who finally blazed an unmistakable trail from asbestos to the grave. A few decades after the war they, and others who had handled asbestos in the manufacturing and construction trades, began dying by the thousands. Even now each year brings to light 2,000 or 3,000 new cases of mesothelioma, in U.S. workers who inhaled the dust long ago. Hundreds more die each year of "asbestosis," a disease that scars the lungs into useless rigidity. And asbestos also causes plain old lung cancer.

And so officials occasionally looked in on W. R. Grace's mine in Libby, Montana. But somehow official concern about how much asbestos the workers were breathing failed to translate into a fear for the whole town. For that matter, it doesn't appear to have provided the workers themselves with admirable protection, either. And even when people in Libby started to be diagnosed with mesothelioma, officials still failed to appreciate the big picture.

In November 1999 the *Seattle Post-Intelligencer* wrote the truth large for Libby: nearly two hundred people had died from asbestos, and nearly four hundred more bore grim diagnoses. The most sobering news was that many of them had never worked at the mine. The whole town seemed to be at risk. The federal EPA rushed to the scene.

"It was the first time anyone pulled together all the pieces," says Paul Peronard, the cleanup coordinator whom the EPA assigned to Libby when the newspaper story broke. To get a handle on the epidemic, Peronard's crew conducted a quick survey of asbestos-related disease in Libby. It looked as though one in five of the asbestos-sickened people had never been employed in the vermiculite business. How had they breathed so much deadly dust?

When Peronard's team dug deeper, they found that many of these victims were family members of vermiculite workers: they were probably doomed by the invisible needles that came home on the workers' clothes. A few of the victims had no family connection. However, many of these recalled playful childhood moments spent scrambling on the stockpiles of vermiculite outside the processing plant next to the Little League fields.

But some victims, Peronard says, seemed to have no direct connection to the tainted vermiculite at all. As more test results came in, an ugly possibility loomed larger: perhaps the entire town had been powdered with asbestos dust. And indeed, preliminary testing turned up the fibers in gardens, in yards, and in house dust. The *New York Times* reported that people had taken home bags of vermiculite for garden mulch, for home insulation, and even to use as a filler in their baked goods. Peronard says he's encountered a recipe in Libby for something called Zonobread, named after W. R. Grace's Zonolite brand of vermiculite.

The EPA quickly warned against home remodeling. "We certainly see asbestos in the insulation. Our preliminary sense is that just having it in the walls is not a big deal," Peronard says. "But disturb it and you could get a con-centration of fibers in the air that would be a concern. If you do that once in your lifetime? I don't know the significance of that. Five times in a lifetime? Daily?"

In fact, despite its obvious deadliness, asbestos dust isn't perfectly under-stood, Peronard says. For starters, the EPA has traditionally considered all strains of asbestos to be equally dangerous in the lungs. But new research, in Peronard's humble opinion, suggests that that view is outdated. Many re-searchers now suspect that curly "chrysotile" asbestos dust—the most com-mon form—is the least deadly. The slimmest, most needle-shaped asbestos, including the "tremolite" and "actinolite" fibers found in Libby, are dozens of times more potent.

Furthermore, health researchers aren't sure how much asbestos dust is re-quired to cause cancer. The Occupational Safety and Health Administration (OSHA) says that the concentration of asbestos in the workplace must be less than two fibers per cubic inch of air. But even at that low level OSHA predicts that one in a hundred workers will get cancer from the dust.

"EPA wouldn't accept a one-in-a-hundred cancer risk for the general pub-lic," Peronard says. "So some people say we need one hundred thousand times less than that. The problem is, we can only detect asbestos to *one* thousand times lower."

Once an asbestos needle slips into to the lungs, the dust can lie low for ten or twenty years before cancer erupts. W. R. Grace closed up shop in Libby in 1990. But nine years later an average week in the bewildered town saw twelve to fourteen new victims sitting through the news that they had asbestosis, lung cancer, or mesothelioma.

Perhaps the biggest difference between the cancerous dust of central Turkey and that of Libby is that Montana's fibers were carried all over the na-tion. An estimated two hundred workers and their families moved away from Libby before the news broke, and they probably carry the dust in their lungs. But more worrying, W. R. Grace sent its vermiculite to about three hundred factories around the country. EPA must locate and investigate every one of them. So far, not so good.

"At first blush the prevalence rates for asbestos-related diseases do seem to be higher in places where these facilities are located," Peronard says. "And again, we have reports of people playing on the piles."

And Libby's special dust may *still* be circulating around the country. The vermiculite insulation, after all, has been poured into the walls of millions of older homes. And it may still be for sale in gardening shops.

In the spring of 2000 the EPA tested a package of vermiculite found for sale in the gardening department of a store in the Seattle area. The *Seattle Post-Intelligencer* once again delivered the news: the Zonolite brand of vermiculite had been packaged at a W. R. Grace factory in California, in 1991. And yes, the vermiculite was contaminated with asbestos. The EPA proceeded to test a large selection of vermiculite gardening products, purchased from all over the nation. Some did contain asbestos, the agency concluded, and consumers would be powerless to tell which these were.

What is the risk of Libby's dust, to, say, a New Jersey gardener who might empty a well-traveled bag of W. R. Grace vermiculite into the tomato patch? The EPA can't say for sure. But at worst it's bound to be much lower than the risk of working with asbestos every day.

W. R. Grace, for its part, has consistently denied it sold products containing unsafe levels of asbestos. The company also says that its mine and processing plants complied with health and safety regulations. But Grace isn't denying there's a demonic dust loose in Libby, and the company is paying the local hospital to screen residents for asbestos-related disease. The lawsuits are piling up just the same.

As industrial dusts go, Libby's was more democratic than average: it touched everybody. That's not always the case with hometown dusts. Often the dusts of industry focus their fury on workers. Although some industrial dust does tend to swirl out into the community, its grip is strongest on those people who suck in the most.

A few centuries ago a European worker who keeled over with his lungs jammed with dust seems to have raised a minimum of scientific interest. Laborers may have been viewed—when they were viewed at all—as expendable. These days workers in the United States are so well protected by antidust regulations that . . . well, actually, thousands of them still keel over every year, their lungs jammed with dust.

And it's not for lack of knowledge. More than three centuries ago a European physician sliced into the lungs of some dead stonecutters and was impressed to discover they were gritty masses of rock dust. And over the following century the connection between a man's workplace and his lung disease slowly revealed itself to curious doctors. The discipline of occupational health began to crawl forward.

Physicians noticed that laborers who spent their days in the company of millstones and grinding wheels contracted a deadly breathing affliction. Revisiting the subject of stonecutters, doctors realized that cutting sandstone was a fatal occupation—but that cutting limestone was not. And when a physician in the mid-1800s analyzed the fine sand he took from the lungs of a dead grinding-wheel worker, science began to close in on one of the most dangerous occupational dusts.

The dust was crystalline silica, better known as quartz. Granite is mostly quartz. Sandstone is made up of small grains of nearly pure quartz. Most beaches are the same. Quartz is the second most common mineral in the Earth's crust, and the list of occupations where quartz dust billowed into workers' faces was a long one. The synonyms for quartz-dust disease show how widely quartz-rich rocks were employed: grinder's rot, stonemason's disease, miner's asthma, potter's rot, rock tuberculosis.

Today at least a million U.S. laborers work with substantial amounts of quartz dust. Each year about 250 of them die with a diagnosis of "silicosis." Masonry workers, miners, rock drillers and crushers, sandblasters, foundry workers, and yes, grinders, are still among the most vulnerable. People who work with "ground silica," a rock powder used in everything from toothpaste to paper coatings, are also at high risk. And to this day pathologists are cutting open dead men and discovering lungs so packed with dust that they cannot be sliced with a scalpel.

That degree of lung dusting is not easy for a mere amateur to achieve. The human lungs are accustomed to hosting a fair amount of dust. They can capture and evict it with dazzling efficiency. Up to a point.

Our lungs take in roughly 14,000 quarts of air each day. In the lungs, oxygen in this air passes through the walls of 500 million alveoli. Their combined surface area, one pathologist has calculated, amounts to a tennis court.

But even moderately grubby air can hold thousands of tiny particles per cubic centimeter. So in the course of one minute's breathing, in moderately grubby air, as many as 30 million particles might stick to the walls of the branching air tubes alone. These are lined with mucous escalators that constantly ship dust out. But in that same minute, another 10 million specks might come to rest inside your alveoli. And those aren't equipped with elevators.

Some of the dusts will dissolve and disperse. But the miniature rocks, the ragged chunks of soot, and other solid dusts will not.

Fortunately, about a dozen cells called macrophages stand guard like bouncers inside each balloon. When a bit of dust touches down, a macrophage will engulf it and carry it to a lymph node or the nearest the mucous escalator. Most of the time these bouncers carry dust out about as fast as it comes in. But the system isn't failsafe. Tobacco smoke, for instance, can slow the mucous escalator. Long asbestos needles can be too awkward for macrophages to handle. An innocuous dust, like iron, can enter in such hordes that it simply accumulates—harmlessly. Or a dust can be so poisonous that it kills the bouncers that try to evict it. And that seems to be the effect of quartz dust.

When a bit of quartz dust lands in the lung, a macrophage will dutifully engulf it. But something about this dust is hard on the bouncer's equivalent of a stomach. The stomach ruptures. The bouncer cell dies. And the dust speck is free again. A second bouncer will ooze to the scene, only to be murdered in the same manner.

The skirmish will end when bandage-making cells weave a fibrous web over the dead bouncers and quartz specks. As more dust, bouncers, and bandages accumulate, a scarlike, dust-studded "nodule" a quarter inch wide will form. As these gritty scars form throughout the lung, each knot reduces the lung's air capacity. Twenty or more years of such dust battles will often produce "classic silicosis," which, on a dark X ray, looks like a sprinkling of fuzzy stars. And while the stiff nodules may cramp lung function, they might not even shorten life.

On the other hand, silicosis may progress to "fibrosis." The bandage makers may weave such a dense web that the lungs can't pass enough oxygen to the blood. People who reach this stage are tethered to an oxygen bottle for the rest of their days. And those days end in suffocation or heart failure.

If a worker labors in especially thick clouds of quartz dust, the disease can speed up. In "accelerated silicosis," the scarring causes shortness of breath after just five to fifteen years. And the worker who breathes *much* too much dust (traditionally, that would include unprotected sandblasters, rock drillers, and ground-silica handlers) risks "acute silicosis" after just two or three years: the traumatized lungs fill with nodules but also drown in a sickly fluid that pours into injured air pockets of the lungs. It is, at least, quick.

But even the horrors of acute silicosis don't seem to satisfy this dust's mean streak. It is increasingly apparent that quartz may promote a raft of diseases, ranging from cancer to immune-system malfunctions. In combination with tobacco smoke especially, quartz dust can cause lung cancer. Tobacco smoke alone, of course, excels at this. But silica dust *alone* may, too.

And lung cancer may be the tip of the iceberg, says David Goldsmith, an epidemiologist at George Washington University in Washington, D.C. Since early in this century, Goldsmith says, an insistent trickle of evidence has linked silica to cancers of the stomach, lymph, and skin, as well as to kidney diseases, tuberculosis, and a host of autoimmune disorders, including rheumatoid arthritis, scleroderma, Sjögren's syndrome, and lupus. That would be a shocking rap sheet, if all the charges were to stick. Just how quartz might do these deeds is still in the speculative stage, Goldsmith says. These potential risks to dust laborers have not yet raised a storm of scientific interest.

"Research on silica issues, up to the 1980s, was considered the backwater of the *backwater* of public health," says Goldsmith, who nonetheless has pursued the subject doggedly. "Now we know a lot more about it than we did, but there's a lot we *still* don't know."

In this day and age it's a bit surprising that people are dying of silicosis. And the United States is not exactly leading the charge against this very preventable dust disease.

European countries severely restricted the use of quartz sand for sandblasting about fifty years ago, for instance. The U.S. government attempted to follow suit in 1974 but was overridden by the painting and sandblasting industries. In 1996 then–Labor Secretary Robert Reich declared a new war on quartz dust. Under the campaign slogan "It's Silica, It's Not Just Dust," Reich implored workers to sprinkle dusty work areas continuously with water, to use a different kind of sand for sandblasting, and to get regular X rays. It was a small war, granted. But many silica workers in developing nations receive an even less spirited defense than that.

"South America, China, Russia," Goldsmith rattles off impatiently. "There are still some very high exposures. In China in 1995 people with silicosis were losing twenty-two years off the average life expectancy. That's a severe problem."

To the public, coal dust is a better-known brute, thanks to the disease known as black lung.

Carbon-rich coal dust doesn't kill off macrophages the way quartz dust does. When a coal speck touches down in the lung, a bouncer cell successfully engulfs it and carries it to a central clearing point. But if the dust isn't cleared fast enough, it piles up. In this case the resulting dust clot is called a "macule," which resembles a black pea wedged into the lung tissue.

Early in the disease a slice of lung will be evenly spotted with these black peas. And regardless of the fact that the miner's sputum will be dyed black with dust, this first stage of black lung isn't a killer. However, of the miners who do reach this coal-spotted stage, the dust will eventually kill one in three.

If a miner keeps breathing dust, his lung tissue will start to fail. Emphysema may degrade the elastic in lung tissue, and the miner will have a hard time exhaling. Or fibrosis, the same scar-making frenzy that stiffens the lungs of silicotics, may set in. A coal worker's lips and ears may turn blue as the transfer of oxygen from his lungs to his blood slows. As fiber continues to bind up the lungs, his breathing will become increasingly labored and painful. Like the silicotic, the breathless miner may spend his final years in the constant company of an oxygen bottle.

Coal dust kills about 1,500 miners each year. It disables thousands more. A black-lung fund, supported by a coal tax, doles out about $1.5 billion each year to 200,000 people—both diseased miners and their bereaved spouses and children.

The evils of coal dust didn't win the attention of U.S. regulators until the 1960s. And now, despite federal limits on coal-mine dust, the death toll stubbornly persists. One reason may be that cheating on the mandatory dust tests inside mines has sometimes been rampant. The federal Department of Labor busted hundreds of mining companies for faking tests in the early 1990s. And in 1998 the *Courier-Journal* in Louisville, Kentucky, reported that the cheating tradition was still in wide practice. In late 2000 the federal government began a takeover of the testing. But even if every mine met the current dust limit, a federal agency that studies miners' health contends that miners would still be risking dust disease. The limit itself is too high, the research agency says.

Those miners who escape black lung may get silicosis instead. Because miners often drill through sandstone and other rocks to reach the coveted coal—or gold, or any other rock-bound treasure—they can suck in plenty of quartz dust. And while surface mines like the expansive open pits in Wyoming would seem to be less dusty than underground mines, they're not. One recent government study of Pennsylvania surface miners found that nearly 7 percent of those who volunteered to be X-rayed had silicosis.

And it's not just workers who inhale the products of mining. In the spirit of sharing with one's neighbors, surface coal mines make a measurable contribution to the dust in the local air. It is not staggering, according to one recent study that compared four towns close to mines with four towns farther away. But the study's authors speculated that this extra dust might explain why children in the closer towns visit the doctor a bit more often than the average child.

Asbestos, quartz, and coal: these are the best known of the industrial dusts that haunt the lungs of U.S. laborers. But nearly any rock or metal dust that a worker inhales for too long will cause its own disease. And while some of these afflictions are becoming quaint footnotes in U.S. medical texts, workers in developing countries are still at risk.

"Talcosis," for instance, is similar to silicosis, but it's caused by talc—ground soapstone. Like ground quartz, talc dust has a hundred uses. Cosmetics, baby powder, pills, paper coatings, and garden fertilizers often contain talcum powder. And to complicate matters, talc deposits are sometimes contaminated with asbestos.

Graphite in the lung behaves a lot like coal dust. One textbook explains that the lungs of a graphite-dust victim "have the appearance of sponges that have been soaked in ink." This crystalline form of carbon is ground up for use in everything from lubricants to pencil "lead."

Potter's lung is caused by feldspar, a silica-rich rock that's milled to make pottery glazes. Whereas the silica in quartz is crystalline, the silica in feldspar is not. Nonetheless, breathing too much of it can fire up the lung's fiber-making cells. "Laundry-worker's pneumoconiosis," first diagnosed in England, may have resulted from shaking pottery-workers' garments before tossing them in the washtub.

"Siderosis" turns up in welders and decorates the lungs with beads of iron or flecks of blackened silver. Sometimes these dusts reside so harmlessly in the lung that they're called "nuisance dusts." But a trickle of evidence links metal dusts to a quick-killing lung disease called "cryptogenic fibrosing alveolitis" and to mouth cancer as well.

Vegetable dusts may seem gentler than metal and mineral powders at first blush. While stone dusts under the microscope often resemble sharp-edged rocks, vegetable dusts look more like soft shreds of rope. But plants have a long tradition of fouling workers' lungs.

When the job of spinning Irish linen thread moved out of homes and into mills in the late 1800s, many workers suddenly developed "Monday morning fever." Workers who toiled in the dustiest parts of the operation suffered a chronic cough, wheezing, and shortness of breath. The symptoms were worst on Monday, after workers had spent a little time away from the mill. Then they would ease as the week wore on. After ten or twenty years of inhaling the dust, the worst-on-Monday disease often became bad full-time. Technically known as "byssinosis," the disease was soon noted in flax, cotton, and hemp mills worldwide.

Like silicosis and black lung, this dust disease persists. As of 1995 OSHA estimated that 35,000 cotton workers in the United States were disabled by a disease that workers call "brown lung." Cotton-dust deaths are rarely recorded in the United States, probably in part because brown lung blends in with other common lung diseases. In poorer cotton-growing nations, however, brown lung runs rampant.

Cotton dust is actually a mix of plant parts, dirt, and bacteria. And those bacteria can be toxic. "Endotoxins" are poisonous chemicals that rise from both living and dead bacteria, to hang in the air like ghosts. As these toxins swarm through a mill, workers inhale them. Still, the exact cause of brown lung remains blurry. Less blurry is the fact that cotton workers who also inhale tobacco smoke do much to improve their probability of developing the disease.

(Synthetic fibers aren't necessarily more wholesome, lung-wise. "Flock-worker's lung" is a brand-new disorder found in some factories that make short snippets of nylon that are used to make pseudovelvet blankets, ribbon, car upholstery, and the like. Increasingly, dusts that scientists once considered biologically inert are showing they can stir up trouble in the lung.)

Wood dust is another vegetable dust with a dark reputation. Many furniture makers, carpenters, sawmill workers, and other wood busters start to cough and lose lung function after just two years on the job. Those who stick it out for ten years—and especially the smokers—stand a good shot at developing asthma. Wood dust has also been linked to cancers of the nose and mouth in a few studies. Some trees make more powerful dust than others, the worst offenders being beech, cedar, oak, mahogany, iroko, and zebrawood. Ripping into wood can also unleash a cloud of molds. And it is those fungal dusts, not the wood itself, that brings on such poetic maladies as "maple-bark-stripper's lung," "sequoiosis," and "paper-mill-worker's lung."

But if the ubiquitous and virulent quartz dust has a sibling in the plant world, that would be the ubiquitous and virulent wheat flour. It's everywhere, and it's potent. To its credit, it seems to be satisfied with causing mere allergies and asthma in the bakery workers who toil inside clouds of it.

Studies show that a whopping 10 percent of new bakery workers may develop a mild allergy to wheat within just one month on the job. After three years one in five bakers shows the same mild result. And after ten years, one study found, fully half of bakers are sensitized to wheat, though only 5 or 10 percent display full-blown asthma.

Like cotton dust, wheat flour keeps unsavory company that may cause some of the trouble. Flour is unavoidably contaminated with everything from wheat hairs to fungi to ground-up bits of weevils, mites, and rodent hair. And each of these dusts can cause asthma by itself. "Wheat weevils disease" describes a powerful allergic reaction to extra-infested flour. The dust of sorghum, barley, corn, soybeans, and oats can inspire the same troubles as wheat. And (needless to say, perhaps) smoking makes matters worse.

Although cotton, wood, and grain dusts are the most famous veggie dusts, the wheezy story continues throughout the vegetable kingdom: people who process tea leaves inhale dusts that constrict their airways. Coffee dust can cause irreversible lung damage. In cigarette factories tobacco workers may suffer something akin to the cotton-dust disease. Cork workers risk an illness caused by moldy cork dust. And then there is—or *was*—"paprika-splitter's lung." This ailment was once the exclusive property of people who split open the fat, red (and moldy) paprika pods in Hungary and Yugoslavia. Selective breeding of the pods has now eliminated the need to split them—and has rubbed out a colorful disease as well.

Mineral, vegetable . . . even *animal* factories can raise a noxious cloud of dust. Everyone knows that the dust of cats and dogs can inspire sneezing allergies and wheezing asthma. But rats? Locusts? Who spends enough time with those camp followers to become sensitized to their dusty by-products?

The people who breed or experiment with laboratory animals do. It seems that mice, rats, and other lab animals often get the last laugh when it comes to dealing out disease: their bodily dusts may give keepers a flulike illness called "rodent-handler's disease," or even a lifelong sentence of asthma.

Airborne particles of rat urine are especially efficient in rousing allergy and asthma in the staff at breeding facilities and labs. And the pee dust has been found inside ventilation components of buildings where rats are kept, suggesting that this stuff may be wafting out into the neighborhood. The airborne specks of urine and saliva of a guinea pig are pernicious, as well. Rabbits and, yes, captive-bred locusts also produce punishing dusts.

Taken together, these small animals will produce symptoms of allergy in about one-third of their human breeders and handlers. If these folks don't take the hint and change careers, the dust will eventually give about half of them asthma symptoms that won't quit—even after the animal handler does.

Let's flee the dusty factories and laboratories, then, and proceed to the farm. Surely, there's no profession more healthful than that of the farmer, who toils in the wholesome sunshine. Even the drive through the clean country air is a pleasure.

On second thought, you might want to keep the windows rolled up. A study from the California Institute of Technology recently found that cars on paved roads kick up a dusty cloud that contains twenty common allergens, including everything from pollen to mold and even animal dander. And a country mile will produce more dust than a city mile, because plants and soil are more common alongside rural roads.

Should you pass that proud symbol of rural independence, the burning barrel, crank the windows *extra* tight. The soot that flies up from this handsome device (a rusty fifty-five-gallon drum) is a signature dust of the rural United States—and of many less-developed parts of the world.

Some 20 million people in the United States still employ the burning barrel to dispose of their household trash, according to the EPA. And the agency surprised even itself when it conducted tests of these primitive incinerators: burning as little as three days' worth of trash in a barrel can produce as much carcinogenic dioxin as a big new waste incinerator does when it burns two hundred tons of garbage. The relatively cool burn of the backyard barrel is key to transforming chlorine into dioxins. And chlorine can be found in everything from bleached paper to polyvinylchloride (PVC) plastic bottles and even table salt.

The bucolic barrels also send forth clouds of smoke—more than a quarter pound of tiny soot particles for each day's worth of garbage. When EPA removed recyclable items from the trash, the burning barrel produced less soot but more dioxin.

But if burning barrels are not in evidence, then perhaps your route to the farm coincides with a garbage-truck route. In that case keep those windows up. The smell of garbage should be a clue that something is escaping from the plastic bags and rising into the air. Garbage collectors probably inhale a bit of everything that the rest of us throw out. Judging from the strains of dung bacteria that dance in the air at garbage facilities, this would include the contents of diapers and bags of dog doo.

And garbage handlers do suffer from this dust. Whether they work in trucks, at incinerators, or even in recycling programs, these laborers report more diarrhea, bronchitis, coughing, asthma, breathlessness, flulike illness, and fatigue than other workers.

But at last, the farm. The good old farm, home to "dust explosions" that can blow apart a concrete silo. The humble farm, where even the cows get "fog fever." The healthful farm, where hearty laborers are routinely floored by a feverish and achy disease called "organic dust toxic syndrome."

ODTS, as the pros call it, is another long-neglected disease that harries blue-collar workers—hundreds of thousands of them each year. Anyone who has ever frolicked in a hayloft knows the throat-thickening sensation that the dust of old hay can produce. That is the gentlest hint of what ODTS feels like. In a 1994 plea for more attention to this agrarian plague, the National Institute for Occupational Safety and Health (NIOSH) recounted a typical case:

> Eleven male workers, aged 15 to 60 years, moved 800 bushels of oats from a poorly ventilated storage bin in Alabama . . . the oats were reported to contain pockets of powdery white dust. Work conditions were described as extremely dusty, and all workers wore single-strap disposable masks while inside the bin. The workers shoveled the oats for eight hours, in groups of two or three, for shifts of twenty to thirty minutes. Within four to twelve hours, all nine who worked inside the bin became ill with fever and chills, chest discomfort, weakness, and fatigue. Eight reported shortness of breath, six had nonproductive coughs, five complained of body aches, and four developed headaches. The two workers who remained outside the storage bin developed no symptoms.

NIOSH cited additional cases, contracted in different pastoral pursuits: A fifty-two-year-old man shoveled composted wood chips and leaves, then twelve hours later entered the emergency room with a fever and difficulty breathing. In another instance, workers who removed a moldy layer of silage from the top of a silo inhaled a white "fog," and they soon fell ill.

That fog isn't pieces of grain, silage, or hay. It's fragments of microbes that live in those products. And if you breathe enough of it, as one-third of all farmers do at some point in their career, a few hours later you'll have an authentic farm experience: ODTS feels a lot like the flu, with breathing difficulties thrown for good measure.

Although the concern about ODTS is new, the disease isn't. In folk medicine terms it's been called "grain fever" and "silo-unloader's syndrome." And thus far the concern about this common fever has not produced an understanding of how the organic dusts do their damage. The culprit is sometimes endotoxins, the toxins that some bacteria release into the air. However, grain can also be contaminated with such malignant dusts as molds, bits of husk, spikes, starch granules, and hairs from the grain itself, plus insect parts, pollens, and rodent hair.

The good news about ODTS is that stricken farmworkers usually recover in a few days. The bad news is that a farmer who has one attack will likely have more. If the ODTS doesn't chase farm workers to the city, they might stay in the business long enough to develop "farmer's lung." This country ailment is an *allergic* condition that can evolve into a lifelong fight against bronchitis, shortness of breath, weight loss, and lung fibrosis that slowly starves the body of oxygen.

Even a cow can get farmer's lung, if she spends too many years with her nose buried in dusty feed. So-called fog fever gives a cow chronic, cow-size coughs. Fog fever isn't usually a debilitating problem for the sedentary cow. But the *horse* version, called "heaves," bodes ill for a racing career.

These vegetable dusts are subtle. The "animal" dusts, on the other hand, announce themselves with unmistakable smells. And their bite may be worse than their bark. Pig dust, for instance, contains everything from mists of manure droplets to dry dung dust, infectious germs, bits of bedding and feed, flecks of pigskin, and a hefty dose of endotoxins. In factory farms, where hogs are jammed into barns by the thousands, the dust is especially hazardous, according to Susanna Von Essen, a lung specialist at the University of Nebraska Medical Center.

"Pig barns have a lot of endotoxin, dust, and ammonia, so there are multiple exposures," she says. "Plus, the profit margins in farming are small, so people don't usually want to spend money to clear the air in the barns." To some extent the power of pig dusts is elusive, Von Essen says. Although endotoxins are an obvious suspect, in laboratory experiments she has removed the endotoxins from dust samples and found that the remaining dusts can still cause cellular swelling. So something else in the dust is capable of injuring lungs.

"There is a 'Factor X,'" she says. "We're working on that." Less mysterious are the viral diseases that float between pigs and people. Drifting droplets occasionally spread meningitis and swine influenza from beast to man, Von Essen says.

Albert Heber is a pig-dust researcher who has firsthand experience in the field. These days the Purdue University professor studies the various germs and other dusts that escape from pig farms to annoy the neighbors. But he spent his youth working on his uncle's pig farm, and he knows how much worse it is for the people inside the barns than it is for those downwind.

"You're always coughing up phlegm," he recalls. "I didn't know what it was coming from. I had that for five years. Then I quit working for my uncle, and it went away. Eventually, I got into this research, and said, 'Wow! *That's* what that was!' Now even when I go into a barn for a few minutes, I can feel that . . . *taste*, those effects, for a day or so afterward."

Heber recalls meeting some farmers in the 1980s who were so allergic to pig dust that they couldn't spend fifteen minutes in their own barns. And although conditions have improved, turnover for employees remains high, Heber says. And it's scant wonder. "The dust we collect on filters is the color of dung," he says. "It's brown. It's fecal dust." In the United Kingdom, Heber says, hog farmers must supply each laborer with a respirator. Not so in the United States.

Poultry farming is no stroll in the park, either. Bird-dust disease was first noted in duck pluckers and parakeet handlers, in 1960. It has since been seen in chicken and turkey farmers. This dust illness is most famous, however, for its association with pigeons. As many as one in five pigeon fanciers will develop an allergic reaction to the birds. And hence it goes by the names "bird-fancier's disease" and "pigeon-breeder's lung."

The dust of fungi in the dry droppings is often nominated as the cause. This line of thought is sometimes criticized because pigeon poo is "hygroscopic": it attracts water and stays gooey. And gooey things aren't prone to disintegrate to dust. "Bloom" is an alternate nominee. Bloom is the truly tiny particles of waxy stuff that pigeons use to waterproof their feathers. And dried serum—blood fluid—that exits the bird on the tip of every fallen feather is also a candidate. All these bird parts contain potential allergens.

As with many of the world's mean-spirited dusts, farming dusts don't stop at the split-rail fence. As surely as a coal mine, a paper mill, or a burning barrel taints the neighborhood air, so does a farm. Pesticides and fertilizers can take a romp through town, as either gases or dusts. Soil, of course, flies around by the ton. Even grain dusts can make a solid contribution to the clean country air. Taken together, the U.S. grain-handling business—elevators and grinding mills—leaks half a million tons of dust into the surrounding air each year, rat hairs and insect bits included. And this dust can pose a danger more catastrophic than disease.

Whenever grain is poured or pushed or tumbled, the kernels grind together and bits break off. This fine dust accumulates in the grain, in the air, on the equipment, and in corners of an elevator. And its tiny size makes it wildly flammable, as workers at the DeBruce Grain, Inc., elevator learned firsthand in 1998.

One of the largest in the world, the Haysville, Kansas, elevator is a row of tall cement silos that march across the plains for a quarter mile. Conveyor belts ferry grain through the complex in underground tunnels. And somewhere in the elevator on the morning of June 8, a match of sorts was struck. A worn bearing in a conveyor belt may have begun to heat up, or a light may have been left in the wrong place by a contractor—the source of the initial spark hasn't been agreed upon.

Whatever the case, some hot item steadily warmed the dust surrounding it.

When a bit of lurking dust finally reached the burning point, a little ball of fire popped open. The fireball expanded as it raced through the dust hanging in the air. This first fireball was weak. But it was strong enough to kick dense clouds of settled dust up into the air. When this newly raised cloud of dust caught fire, a second explosion blasted car-size chunks of concrete out of the silos. It crumpled steel doors as though they were aluminum foil. Houses shook in Wichita, ten miles away. Dust, grain, and injured workers were everywhere. Seven men died, and four of them were buried in the grain for days. The grain itself burned for weeks, leaking smoke out of the ragged elevators.

And Haysville was not alone. That year produced a bumper crop of eighteen dust explosions in the United States. These blasts injured a total of twenty-four people and ruined property to the tune of about $30 million.

But in terms of a farm dust's impact on the neighborhood, even a dust explosion may not be as problematic as the dust of, say, a few thousand hogs. As far as the neighbors are concerned, vegetable dusts are a piece of cake. Airborne animal dusts, however, tend to announce every shift in the wind with unmistakable odors.

At this stage of his career Professor Albert Heber spends his time *outside* the hog barns, finding ways to minimize the noxious dust that reaches the noses of neighbors. One of his studies found that so many bacteria were escaping from a pig barn that a plume of them could be measured more than two hundred yards downwind. Bacteria were still detectable—and alive, in spite of dry, sunny weather—a football field away. Bacteria aside, the pig dust itself can be thick enough to stink a *mile* downwind.

"People don't usually complain about dust," Heber says. "They complain about odor. But dust *carries* the odor. They just can't see the dust."

While pig dust is legendary for its stench, cow dust probably dominates in terms of tonnage. Feedlots, where cattle are fattened for the slaughter, are especially productive of what the experts call "fecal dust." Each day that a beef animal spends fattening in a pen, it produces five or six pounds of manure—dry weight. If the pens aren't kept damp, the cow pies soon lose their water content to the air. Then, in the cool of the evening, the cattle grow restless. They give the dry cow-pies a good stomping, breaking them into tiny particles. And away go the pulverized pies.

Do they ever go. In one day a thousand corralled cows will send about fifteen pounds of pie dust stampeding through the neighborhood. The cumulative effect can be gruesome. In Texas on any given day there are usually about 3 million cattle churning out the pies in feedlots. Nationwide, livestock kick up something like 65,000 tons of dust each year. Compared to the tonnage of eroded cropland that takes to the wind, that's a minor puff. But to the neighbors who wipe it off their dinner tables, perhaps the special qualities of animal dusts compensate for their relative rarity.

Dusts of farming and industry are easy to spot, and easy to blame, when a plague of death or disease surrounds them. But some of nature's own dusts are perilous, too—as the case of the Turkish cave towns proves. Depending on where you live, your local dust may be a monster. Just ask an Egyptian mummy.

In 1973 a medical team at Wayne State University School of Medicine in Detroit brushed aside the risk of "mummy-unwrapper's lung" (it truly exists) and proceeded to autopsy an Egyptian man who died two thousand years ago. They discovered all sorts of fascinating things in his body, including a chunk of worm-infested meat in his intestine and sclerosis in his arteries—a condition we attribute to our modern fatty diet. But it was his lungs that told doctors about the ancient Egyptian air: it was dusty.

Patches of his lungs were fiber-stiffened. And in those diseased areas were little deposits of two types of dust: one was black carbon, from smoke. The investigators noted that this sort of lung condition was predictable and "probably occurred as soon as man made fires in confined spaces, such as huts, caves, or tents."

But the other dust was quartz. The mummy had silicosis. Examining the mummy's hands, the team concluded that this Egyptian had not been a quarry worker or some other manual laborer. There was only one way he could have acquired his quartz dust. He had simply lived through enough desert dust storms to collect a lungful.

(Another dust that had incorporated itself into his bones told a happier story about the air of ancient Egypt. The concentration of lead in his skeleton was less than one part per million. Modern bones hold six to twenty times more of the toxic metal, which is released into the air from smelters and cars that still burn leaded gasoline and from blowing soil, which recycles old pollutants.)

The mummy's affliction, "desert silicosis," is still a plague for some desert-dwelling people. Scientists first reported this dust disease in people of Libya and in bedouins of the Negev Desert. It has since cropped up in some surprising places.

In 1991 an Indian and English medical team announced that it had discovered clusters of silicosis two miles up in the Himalayas of northern India. The village of Chuchot, in a high river valley, endures springtime dust storms thick enough to blot out the mountain scenery. The village of Stok, which sits higher, gets a bit less dust. But both towns get enough dust to plant the stars of silicosis in the lungs of their citizens. The team's X-ray survey found that nearly every woman in Chuchot had some degree of silicosis, as did more than half the men. (The women, the researchers noted, do most of the farmwork in the dusty soil, and they are also responsible for cleaning the dirt-floored houses.) The silicosis rates were a bit lower in Stok, but they were still fantastically high. And in a third village the same researchers later found three people suffering very advanced silicosis.

Desert silicosis has also turned up in farmers of California's Central Valley, in zoo animals kept in semidesert San Diego, and in horses on California's Monterey-Carmel peninsula. Some scientists, however, argue that desert silicosis is often a misnomer. The lungs of desert dwellers do hold quartz dust, and sometimes they also show the fibrous scarring that typifies quartz overload. But those lungs also tend to be rich in the black soot of indoor cooking fires. After studying South African women who grind corn with sandstone tools, a pair of researchers proposed that "hut lung" is a better name for the smoke-and-dust disease of desert people.

Like industrial dusts, natural dusts also come in both "mineral" and "vegetable." Pollen is probably the most widely recognized little nuisance worldwide. But some vegetable dusts are specific to a place.

"Valley fever" is caused when a particular dry-soil fungus finds its way into human lungs. Named for the San Joaquin Valley, this fungus infects thousands of people throughout the southwestern United States in a year, when it blows into their faces uninvited or when they stir up dry soil. Also called "desert fever," or "desert rheumatism," the fungus causes lung congestion, fever, and aches, but it's rarely fatal. It gets little press. And in Simi Valley, California, where this fungus infects just a few people each year, Valley fever is hardly known at all. At least, it wasn't before the Northridge earthquakes struck the hills north of Los Angeles in 1994. But soon afterward more than two hundred people in the Simi Valley area suddenly came down with the fungal fever.

The Centers for Disease Control sent epidemiologists Eileen Schneider and Rana Hajjeh to the scene to figure out where the fungus dust had come from. The investigators were met with news that the earthquake and its aftershocks had launched thousands of landslides. As the dry soil peeled loose from hill-sides, clouds of dust rose into the air.

"These clouds were impressive," Schneider says. "There are pictures of clouds covering whole towns, almost. And then some were only as big as a few houses. And people said the dust settled everywhere—in pools, houses. People wore makeshift masks, trying to protect themselves."

It took a few days for victims to begin showing up at emergency rooms. They complained of coughing, fatigue, night sweats, and fever. Many were mistakenly diagnosed with pneumonia. Most recovered. In fact, Schneider and Hajjeh found that three-quarters of those who became infected were never hospitalized. But three victims died.

When the team mapped the location of the victims and the paths of the dust clouds that rose from the various landslides, they found a match: dust clouds had drifted through the same neighborhoods where people got sick. Cementing matters, the investigators found that those people who reported spending more time enveloped in a dust cloud were likelier to have developed Valley fever. Like many dangerous dusts, this one is poorly understood.

"No one really knows how many spores you need to inhale in order to de-velop symptoms," Schneider says. "Some people think you only need one spore. Others say you need more."

Even the lifestyle of the flying fungus is mysterious. You can collect one soil sample and find spores, Schneider says, then walk fifty feet and collect another sample—and find none.

Hantavirus is another airborne menace that turns up in selected spots. But to keep things interesting, those spots shift around the face of the Earth from year to year. When the disease was diagnosed in 1993, it appeared to be a spe-cialty of the southwestern United States. But the CDC quickly found that this virus was just part of a well-known clan of nasty bugs spread by rodents world-wide. And they realized it could strike anywhere, depending on the weather. All it takes is a bunch of infected mice.

In 1993 the Southwest had extra rainfall and plentiful vegetation. The mouse population boomed. Hantavirus, which an infected mouse excretes in its dung and urine, boomed, too. Then, when a part of Panama received extra rain at the end of 1999, hantavirus popped up there. To keep the CDC on its toes, in 2000 the California Health Department announced that another mouse-dust germ, this one called arenavirus, had killed at least three people in the previous fourteen months.

In order to rise into the air where people can breathe it in, hantavirus and arenavirus do require a degree of dryness. But even an old cabin can shelter the wastes as they dry to dust. Then, when a camper or desert wanderer stirs up that dust, the virus dives into his lungs. Hospitals can nurse a patient through a hantavirus infection. But it is often fatal simply because victims' lungs fill with fluid so fast that they never make it to a hospital.

Yet another natural hazard is the wily puffball. The puffball mushroom is a spherical fungus that occurs in many parts of the world. Various species can be an inch, or several inches, in diameter. As a puffball ages, it fills with dark spores. In springtime these "puff" out of holes in the mushroom's leathery skin and strike out to start the next generation. Unless someone snorts them.

In some cultures, folk medicine prescribes a snort of puffball spores to stanch a nosebleed. Presumably, the ripe spore ball is to be aimed into the nostril and squeezed. But perhaps the secret to keeping the spores from traveling through the nose and reaching the lungs has been lost in the mists of time. When spores do reach the lungs in sufficient numbers, they can cause the patient to forget all about the minor inconvenience of a nosebleed. The side effects of the nosebleed cure resemble a robust case of pneumonia, combining fever, nausea, headache, and breathlessness.

The CDC published a refresher lesson on the puffball illness in 1994, after southeastern Wisconsin puffballs lured a flock of teenagers into their midst. As if the puffball's nosebleed-quenching powers weren't attractive enough, one of the species is also psychoactive, the CDC explained. And so at a party that spring, eight teens with recreational intentions snorted and chewed on puffballs. A spree of vomiting ensued, followed a few days later by a stay in the hospital when the long pneumonialike phase set in.

It was not, perhaps, the trip they'd had in mind.

Human beings invented neither dust nor stinky dust nor deadly dust. Even before people were scattering their own dusts over the surface of the planet, no two spots on Earth boasted quite the same assortment of specks. But in many of the places where we have set up shop, we have embellished the natural themes—until we have all but drowned them out.

In a few spots the original dusts still dominate. In some deserts, at some beaches, even in some Turkish caves, you can still catch a whiff of the local dust, fresh and unadorned. Along the unpeopled coast of Maine, granite dust mixes with flecks of sea salt in the air. The warm pine trees gripping the rock send out sticky blobs of pinene. A red squirrel tearing into a pinecone launches a little brown cloud of cone dust. Dark and oily otter droppings release reeking bits of digested fish to the wind. These dusts give that place a smell all its own.

And when you turn away from such a wild and dusty spot and head back into the cluttered air of modern life, you'll carry souvenirs in your chest. The sea salt will dissolve and get lost in the ocean of your body. The pine dust might as well. But perhaps a few of those specks, the hardest and the most ruthless, will be stowed in a corner of your lung, yours to keep for life.

10

MICROSCOPIC MONSTERS
AND OTHER INDOOR DEVILS

Although rare diseases like ebola, West Nile, and even dusty hantavirus make thrilling headlines, under our noses a much broader epidemic is undergoing a subtle explosion: asthma.

The number of people affected has grown by roughly 50 percent each decade since about 1970. Many of the new victims are children, among whom the asthma rate has doubled since 1980. A total of 13 million people in the United States now suffer from the wheezing disease. They make 2 million trips to the emergency room each year, and each day fourteen die of suffocation when their breathing tubes swell shut. Should the spread of asthma in the United States continue, by the year 2020 about one in ten people will be afflicted. And this explosion is rocking other wealthy nations, too.

"New Zealand, the UK, the Netherlands, Japan, Australia," marvels University of Virginia asthma researcher Thomas Platts-Mills, ticking his fingers: all have skyrocketing asthma rates. Finland is especially unbelievable. Its military records show that the number of nineteen-year-old men with asthma has multiplied *twenty times* since 1960. It is particularly pronounced in cities.

Before this explosion the best predictor of who would get asthma was genetics: it runs in families. But heredity can't possibly explain the huge numbers of new victims. What, then, is the second-best indication of who will get the wheezing disease? It's an allergy: people who are allergic to some component of house dust have a higher risk of suffering from asthma. And indeed the number of people with dust *allergies*—whether it's to dust mites, molds, cockroaches, or pet dander—is also soaring. Our relationship to dust has changed in some crucial way, says Platts-Mills, who sits in his office beneath the gaze of a beige-velvet dust mite toy perched in a bookcase.

"In the 1980s," Platts-Mills says, "we assumed the cause was tight houses, more heat, more carpeting, more furniture." In other words, more places for

dust and mites to accumulate and less air circulation to flush away the allergens. But the asthma explosion has forced a rethink, and Platts-Mills has a
new theory. House dust must bear *some* of the blame, he believes, but not all of
it. Noting that asthma targets rich nations, he proposes that part of the blame
lies in the way the well-off use their lungs. Those organs aren't getting enough
exercise to stay strong when they're confronted with an extra dose of dust.

But his isn't the only explanation that's being put forth. An opposing theory, the "hygiene hypothesis," suggests that old-fashioned house dusts—germ-
ridden dusts in particular—actually built stronger children. The *un*dusty
nature of modern life is putting children in peril, says the hygiene hypothesis.

However the relationship between dust and our lungs has changed, scientists are now putting house dust, long dismissed as a nuisance, under the microscope. What they're finding is often frightening—and the discoveries often
reach beyond asthma.

When researchers venture into someone's house to analyze dust, they often
start by looking at what's floating in the air. That, after all, is the dust you are
most likely to inhale. In the course of their work these scholars have discovered
something they now call "the personal cloud." It accounts for a lot of the dust
we breathe every day. And it's not entirely clear what it's made of.

In a landmark experiment with house dust in 1990, researchers wired 178
people in Riverside, California, with personal dust monitors. The participants
wore their monitors for twelve hours at a time, as they cooked, read, slept, and
so on. During the experiment the researchers also monitored the air inside and
outside the homes for dust. The results were surprising.

Both indoor and outdoor dust levels in this particular neighborhood were
fairly high—half to two-thirds the federal limit, says Lance Wallace, an environmental scientist with the EPA and one of the investigators. But the monitors people wore on their bodies rang up even *higher* levels of dust. If a human
being could be given a ticket for emitting too much dust, this study found that
about one in four of the citizens of Riverside would be reaching for their wal-
lets. "The personal cloud accounts for about one-third of a person's total dust
exposure," Wallace says. "It's a *big* source." And it's something of a mystery.

Skin was the biggest recognizable element of the personal cloud, Wallace
says. When scientists examined the filters from the personal monitors, they
found that a twelve-hour shift sent 150,000 to 200,000 skin flakes into the
dust monitor's inlet.

That's a tiny fraction of the skin an adult sheds. In one day an adult may lose about 50 million scales, which under the microscope bear a resemblance to a wind-scattered newspaper. Many of the day's tiny scales presumably go down the bathtub drain. Wallace guesstimates that you breathe about 700,000 of your own skin flakes each day. And the rest sink slowly to the floor—or entangle themselves in the fibers of your sheets, or work their way into the couch cushions, or accumulate on the lamp shade. Could breathing your own skin give you asthma? It's unlikely. However, analysis of the Riverside dust filters showed that skin amounted to no more than about 10 percent of the personal cloud. It was only the biggest *recognizable* breed of dust on the filter, says Wallace.

Lint would seem like an obvious contributor to the personal cloud. The soft mat of fibers on the clothes-dryer filter gives some indication of how easily your clothes give up their fibers. Turn your head and the friction between neck and collar tears at the fabric, releasing a puff of microscopic fiber. Cross your legs and more fibers are scuffed off. But, says Wallace, lint didn't turn up on the filters in nearly the same quantities as did skin flecks. The majority of the dusts in the personal cloud are still unknown.

Wallace may be closing in on this little mystery. In an experiment funded by the EPA, Wallace wired his own house in Reston, Virginia, with a fleet of dust and gas monitors. His biggest eye-openers, he reports, came in from outdoors: his indoor gas meters tip him off when the neighbors are stoking their fireplaces. And each weekday morning a wave of automotive by-products rolls through the house, although the highway is over a mile away. The extraordinary cloud of soot generated by a citronella candle was also memorable. But one day Wallace accidentally waved an arm toward a dust monitor and noticed that the instrument recorded a flurry of dust.

"I started waving my arms in front of it, and I got a reproducible effect," he recalls. "I called a colleague who had the same monitor in his house and said, 'Wayne, wave your arms in front of the monitor!' And he got no effect. Finally, I asked, 'How do you launder your shirts?'" Colleague Wayne, it seemed, sends his shirts out. His cleaned shirts sit wrapped in plastic until the day he puts them on. Wallace's home-laundered shirts, by contrast, hang in the closet for days before being worn.

"Washing will get rid of particles on a shirt, but more particles will just settle on them again," Wallace concludes. "We now have some data showing that our clothes collect dust from the surrounding air and probably reach some sort of saturation point within a day of being washed. And from then on, the millions of particles striking the clothes either bounce off or stick and knock other particles off. So maybe the personal cloud is just being shaken off your clothes." While vigorous motion creates the biggest personal cloud, Wallace found that even working quietly at a computer in his home office multiplied the amount of dust in that room by five times. And if this is just a sample of mixed house dusts, then the personal cloud might qualify as an asthma suspect—if only because house dust *always* contains various allergens.

But the personal cloud must also be spiced with those dusts we jump into with both feet: Woodworkers will have a woody cloud, and knitters will have a woolly cloud. Cooks will trail flour, and cleaners will disperse laundry powder. Romantics? Candle and incense soot. Beauties? Eye shadow. Gardeners? Pesticide, fertilizer, potting soil. For better or worse, our personal clouds are what make our houses smell like home.

If you already suffer from asthma or allergies, the last thing you should do is try to clean up your house dust. Housecleaning has always been an effective method for moving dust from the floor to the air, where you can inhale it. It is a perversely dusty undertaking.

The humble broom, for instance, moves dirt around by flicking it forward with a certain degree of violence. While the heavier bits of detritus skid along the floor or make short hops, the lighter dusts take wing. Few household activities can compete with this method of dust raising. It is no wonder the vacuum cleaner was such a hit.

The first vacuum cleaners, produced at the turn of the last century, required one person to pump and a second operator to maneuver the dust-sucking hose. After World War I the technology shrank to a more manageable, and affordable, configuration. The broom was on the run. But recently, it came to light that the vacuum cleaner had a dirty little secret of its own: it sucked up dust to beat the band, but in some machines, the smallest, most breathable specks whistled right through the filter and billowed out into the air.

Michael Hilton designed a testing program for vacuum cleaners during his tenure at the Carpet and Rug Institute (CRI), an industry group in Dalton, Georgia, that represents carpet manufacturers. When Hilton initially tested the dust-leaking rates of thirty randomly chosen machines, the worst offender was a shop vacuum that blew out ten times more dust than hangs in the air of a big city on a bad air day. (The comparison is a bit slanted, since Hilton included all sizes of dust that exited the vacuum, while federal standards apply only to dusts less than a tenth of a hair's width in diameter.) A handful of the thirty vacuums, Hilton says, produced about half the concentration of dust that federal law allows in outdoor air. And some leaked an impressively small amount of dust.

The revelation that some vacuum cleaners could increase the amount of dust in the air has led to a revolution in vacuum cleaners. In the late 1990s new models began to tout their HEPA (high-efficiency particulate air) filters. HEPA filters do well at catching small particles that can penetrate the lungs. But this "advance" actually gave some vacuum cleaners *another* dirty little secret.

"We took one vacuum, and put a HEPA-rated bag on it," recalls Hilton, who has since taken a job with a carpet maker. "It reduced the airflow. And that reduced the amount of soil the vacuum removed from the carpet. So HEPA vacuums filter almost one hundred percent of what they take in," he concludes with a chuckle, "which may not be much." And non-HEPA vacuums, he adds, can do a perfectly acceptable job of filtering dust-mite allergens, if that's your goal.

CRI has set its standards, which are completely voluntary. To pass, a vacuum must emit less than two-thirds of the federal outdoor-dust limit. It must remove "a satisfactory quantity" of test dirt from a carpet in just four passes. And it mustn't inflict visible damage on the carpet. When Hilton was developing the standard, fewer than 30 percent of the vacuums he tested passed. And not much has changed since then, he reports.

So vacuum cleaners can raise dust that may contain allergens, which could aggravate an existing case of asthma. But if vacuums throw up less dust than old-fashioned brooms, this kind of cleaning can hardly be blamed for *increasing* the prevalence of asthma.

Perhaps a more recent development in cleaning is changing our relationship with dust. It has been suggested that houses got dirtier when women started going out to work. Perhaps dustier, smellier houses were the reason that "air fresheners" were able to invade the family shopping bag so easily.

Few names could be more misleading. Some of these products do include a chemical that might bind with a drifting malodorous molecule, making it heavier and more prone to stick to something. But otherwise air fresheners often "clean" in one of three ways, according to an expert at the EPA. They numb the nerve endings inside your nose, preventing you from smelling anything offensive. Or they coat the inside of your nose with droplets of oil, to the same effect. Or they simply saturate the household air with such strong-smelling chemicals that other odors fade by comparison.

Whether these chemicals spray out of a can, waft from a disk, or boil from an oil, some of the compounds will settle in for a long stay. Those ingredients that prefer a gaseous state might eventually waft through cracks in the wall and head outdoors. But those ingredients that prefer a gooey or solid state will drift for a while as wee redolent particles, then stick to walls, cling to cobwebs, and sink into carpets.

In the 1990s a new pseudocleaning product challenged air fresheners for the right to mask household mildew, smoke, and other stinky dusts. Candles had a more wholesome, old-fashioned image, and with a slug of perfume stirred into the wax they could camouflage unpleasant smells. Scented candles have been around for decades, but people used them mainly to set a mood. In recent years two new trends have kicked scented-candle sales through the roof. One is the marketing of candles as air fresheners. The other is a New Age take on mood setting, whereby perfumed candles deliver the psychological benefits of "aromatherapy." As a result, in the early 1990s candle sales grew a raging 10 or 15 percent a year. In the *late* 1990s they doubled that growth rate. And who would suspect all these candles, these symbols of simpler times, of making toxic or destructive dust?

Public Citizen would. The consumer-protection group had a hunch that the candle craze was spreading lead dust through people's homes. So from stores in the Baltimore and Washington, D.C., areas, Public Citizen investigators purchased about ninety candles, choosing those whose wicks held a slim metal core. In February 2000 they announced that their suspicions were justified: 10 percent of the wire cores contained a substantial amount of lead.

When you burn metal wire, lead or otherwise, the metal essentially boils off into the air. And there it quickly cools into tiny specks of dust. Public Citizen reported that burning a candle with lead in its wick for three hours a day in a fifteen-by-fifteen-foot room put enough lead in the air to blow the doors off federal standards. The group has called for a ban on the wicks.

(An easy home test is to attempt to "write" with the end of a candle wick. Lead will leave a gray smudge. The safest bet, Public Citizen concluded, is to avoid wire-core wicks altogether.)

Lead dust is a bit player, however, compared to the old-fashioned soot that some candles produce. Consider a thoroughly modern phenomenon called "ghosting." Frank Vigil, a building-science specialist at Advanced Energy Corporation in Raleigh, North Carolina, was among the first people to notice this spooky development: black lines and smudges are appearing on walls and carpets. They're soot, deposited by air currents. And Vigil thinks candles are a key contributor.

"Many of them have a higher level of aromatics—perfume. And aromatics, we believe, sometimes lead to less-complete combustion. And that makes soot. Take jar candles: The reason they're in a jar is often because they're high in fragrance, and the wax is softer. And because they're inside a jar, they starve for oxygen." And that makes soot, too.

Operator error worsens the problem, Vigil continues. "People don't trim wicks properly. It should be one-quarter to three-eights of an inch long. But that would mean you burn a candle for about two hours, blow it out, and trim the wick. Most people don't do that. People burn candles in front of drafts. That causes flickering, and flickering makes more particulates. Just blowing out a candle produces tremendous particulates. You should bend the wick down into the wax." Candles are not the sole ghost makers by a long shot, though.

"Furnaces, water heaters, fireplaces, oil lamps, space heaters—you can get soot just from toasting an English muffin," says Vigil.

So pseudocleaning can certainly thicken the dust in a house. But if there's an asthma instigator hidden in the ingredients of these products, researchers haven't isolated it.

Even the daily cleaning of bodies can be a source of troublesome dust, although the health risk is not for asthma. Talcum powder is much loved because the rock it's made from is naturally slippery, which reduces the friction of skin on skin. It's also naturally absorbent—a rare quality in a rock.

But talcum powder is, in the final analysis, powdered stone. And human bodies aren't designed to breathe stone dust—or to absorb it any other way, either. Believe it or not, people do perish from baby-powder overdose. The clearest cases are the smallest: more than five thousand times a year a poison-control center in the United States receives a call to report that a baby has been accidentally clouded in powder. Only about three hundred of these infants require medical treatment for their resulting respiratory symptoms. But occasionally, these powder accidents do kill babies.

The ill effects of breathing clouds of baby powder are so reliable that at least one adult has done it intentionally. Munchausen syndrome is a mental illness whose victims feel compelled to cause their own disease. The medical literature describes a case in which a respiratory technologist with Munchausen syndrome used talc to give herself the symptoms of asthma. When confronted with the peculiar results of her lung biopsy, she confessed that she had taken to inhaling the hospital's own baby powder. Though the particles were three to ten times larger than the dusts that doctors normally expect to enter the lungs, they had entered in droves.

When women shake this same dust into their underwear to absorb moisture, it may cause a more dire disease. "Talc has been looked at for ten years now," reports Bernard Harlow, an epidemiologist at Brigham and Women's Hospital in Boston. "The results seem very consistent: talc is probably a risk factor for epithelial ovarian cancer. It's probably causal. But it probably accounts for a small proportion of the total cases." For now the link between talc dust and this common, deadly cancer is circumstantial: women who powder their underwear for at least twenty years are more likely to develop ovarian cancer than women who don't use talc, according to one of the recent studies. Harlow and his fellow talc researchers feel confident that talc accounts for about 10 percent of these deadly cancers. But he admits more research is needed.

"It's not well studied," Harlow says with some frustration. "There's no outcry that this is a big problem like AIDS or breast cancer. And maybe it's not. But what bothers me the most is that it's so *avoidable*. There really is no reason why women have to use it." Cornstarch-based baby powder, he notes, has shown zero correlation with cancer.

Now, the eerie part of this tale is that talc bears a strong chemical resemblance to asbestos—which definitely causes cancer. In fact, the two rocks sometimes intermingle in the planet's crust. And in the 1980s researchers found asbestos fibers in some samples of talcum powder. For the record, that household name in baby powder, Johnson & Johnson, maintains that its brand of talc never contained asbestos and that careful mining practices continue to protect the purity of its powdered mineral. But the specter of asbestos contamination continues to haunt talcum powder.

Whatever you puff, and wherever you puff it, some of the powder evades your sticky skin and rambles freely through the house. Cornstarch powder especially can constitute a bounteous buffet for the critters that live in house dust. The drifting starch granules collect in cracks and have even been noted in the ducts that distribute air through some houses. From the humble perspective of molds, mites, and other dust-dwelling creatures, a morning puff of cornstarch powder must arrive on the air currents like a shower of breakfast cereal.

Cooking has always been another of the dustiest duties in the house—and in the cave or tent. In the lungs of some bedouins and others who cook over an open fire, doctors routinely find rich black deposits of soot. And the modern attempts to keep fire inside a box haven't always squelched the dust.

Some of the most murderous household dusts fly up from Chinese coal stoves. In both cities and rural areas simple coal stoves fill houses with so much soot and sulfur that the notoriously dirty *outdoor* air pales in comparison. Keith Florig, a research engineer at Carnegie Mellon University, has studied China's massive pollution problem. He says that even in cities a terrifically popular item is a small stove that holds a few coal-powder briquettes. The stove can easily be carried from room to room—which means it has no chimney. While the briquettes release less smoke than does raw coal, what smoke they do produce remains indoors.

The death toll from this smoldering coal is staggering. For instance, the lung-disease rate among people sixty or older is twenty-five times the U.S. rate, says Florig. And frying food on a coal stove is a particularly unhealthy chore. Florig cites a study showing that Chinese women who fry are *nine times* more likely to develop lung cancer than are their comrades who prefer other cooking methods. As the fryers stir food, their faces are in line for an upward shower of stove soots *and* airborne specks of cooking oils rendered poisonous by the heat. Frying is, at least, less dangerous than *sleeping* with a coal stove: Chinese women who heat the bedroom with a coal stove for thirty years, another study reports, are *eighteen* times more likely to develop lung cancer than those who never fire up the black rocks in the bedroom. "Lung-cancer rates in Chinese women are about the same as in U.S. women," Florig says. "But in the U.S. it's almost all tied to cigarette smoking. There's very little cigarette smoking among Chinese women." And for good measure, some coal in southwestern China imparts a substantial side order of poisonous fluoride or arsenic to the household air—and then to food. Both can cause terrible disease and disfigurements.

Coal is an especially dusty fuel. But wood stoves and fireplaces also leak soot into the house. And as with coal, study after study has linked indoor wood burning to respiratory disease, especially in children. In the United States these soot-rich fuels are now uncommon. Woodstoves and fireplaces are likelier to smoke in the name of ambiance than for survival. So these dustmakers make poor asthma suspects. However, even we whose kitchens contain smokeless gas or electric stoves have some cooking dusts to consider. The food itself produces enormous clouds of dust.

Eileen Abt is an indoor-air researcher who in 1996 wired four Boston-area homes with dust monitors to gather data for her doctoral thesis at Harvard. During the days that the meters ran, Abt had her human guinea pigs keep a diary of their activities: cooking, vacuuming, chasing children. And when Abt compared the two sets of information—activity and dustiness—she found out what makes the most dust in the house. Just moving around, taking the personal cloud for a walk, was a big dust raiser. But cooking shifted dustmaking into overdrive.

In a graph of her dust data Abt has marked the time that one of her families turned on the oven. The number of supersmall particles trickling through the dust meter remains unchanged for about fifteen minutes. Then the dust count shoots nearly straight up. By the time dinner is cooked, there are about twenty times more supertiny particles swirling through the house. And when the oven is turned off, the line on the dust meter swoops back down. Abt found that baking, broiling, toasting, barbecuing, and frying all produced prodigious quantities of these minuscule motes—dusts a few thousandths of a hair's width in diameter, or smaller.

Sautéing, on the other hand, sent much thicker clouds of much *bigger* particles through Abt's meters. On the "sautéing graph" the flame leaps under the pan in a house at eight-thirty one evening. Within fifteen minutes the number of particles in the air has hit more than *four hundred* times the presautéing level.

And what are these cooking dusts? They're tiny flecks and condensed vapors of whatever's cooking—altered by heat. They can be nasty indeed. When the muscle tissue of vertebrate animals (animals with a backbone) are cooked either too hot or too long, some of the most powerful mutagens known can be produced. Mutagens damage DNA, which may lead to cancer. Most of these heat-forged mutagens stick with the food, and you eat them. They're linked with cancers of the stomach, gut, and breast. But some of the overheated meat chemicals leap out of the pan. The soot of burning meat that rises into the air holds some thirty-four different compounds, including both mutagens and known carcinogens, according to frying studies at Lawrence Livermore National Laboratory in Livermore, California.

Like meat, some vegetable oils can also turn to mutagenic smoke if they're heated too hot, according to research at the National Cancer Institute. Far and away the worst oil, mutagen-wise, is rapeseed, the unrefined version of canola. It is popular in China, which may help to explain why frying food there is so hazardous. But many smokes are dangerous in one way or another. Even burning toast can produce mutagenic soot.

When all these cooking particles first leap up in the kitchen, they're tiny. Then, as they drift through the house, they may merge with one another and grow. Some are inhaled. Others will eventually come to rest just about everywhere: on the wall, a painting, curtains, the floor.

Cooking produces dry dusts, too, of course. Many of the powders in the kitchen cabinet are too big to travel far on the household's weak wind. But the cook herself will distribute cocoa and flour when she brushes off her hands or when she takes her personal cloud for a stroll through the house. Though these dry dusts are innocuous in comparison to mutagenic smokes, they're essential to the health of your house dust. For one thing, these plant powders feed a hungry crowd of creatures waiting on the floor. Bacteria, fungi, silverfish, dust lice, and other scavengers all adore a sloppy cook. Even something as enormous as a fruit fly can eke out a living from fallen food dust. The larvae of these winged wonders have been known to survive on fermenting wheat flour that collects in cracks.

Furthermore, these powders add a diversity to your dust. Grains and spices are impossible for growers and processors to keep clean, and so federal regulations spell out how many pieces of insects and other freeloaders may inhabit any given powder. A cup of flour, for instance, isn't legally filthy until it contains about 150 insect fragments and a couple of rodent hairs. Cornmeal, cocoa, and other powders must meet similar guidelines. These little shards of dead animals contribute to your cooking, and your air, and your dust bunnies.

Besides smokes and powders, cooking can give rise to a third category of dusts, those created in chemical reactions. In one Boston-area home Eileen Abt's monitors may have caught these dusts just as they popped into existence. "If there's ozone in the air, and someone cuts an orange, there's a reaction that occurs between that ozone and limonene in the orange: it produces ultrafine particles," Abt says. "Someone was cutting oranges one day, and I think I saw that happen."

Limonene is a common ingredient in citrus-scented household products and ozone is an important ingredient of smog. So regardless of our culinary abilities, most of us have probably cooked up a few servings of these minuscule dusts and sent them wandering through the house.

Is cooking causing asthma? The soots and chemicals are a diverse, and sometimes dangerous, bunch. But human beings have been inhaling the smoke of burned food for a long time. And the asthma explosion began just a few decades ago.

Perhaps our other appliances are the culprits. Compared to cooking and cleaning, humidifiers, hot tubs, and other watery appliances are subtle. But they have all been known to quietly toss out handfuls of living and dangerous dusts.

Legionnaires' disease is caused by a bacterium that usually enters the lungs hidden in microscopic droplets of water. It is so named because it was first described in a group of veterans who had gathered at a Philadelphia hotel for a convention in 1976. Since then clusters of cases have often been traced to building-top cooling towers, where warm water allows the bacterium, *Legionella,* to multiply.

But the federal Centers for Disease Control estimates that between 8,000 and 18,000 people in the United States develop Legionnaires' disease each year—and they don't all get the bug in big buildings with cooling towers on the roof. Many of these cases probably originate from water-flinging appliances at home, although CDC says some victims also inhale the bug from potting-soil mixtures.

Normally, this bacterium prefers to reproduce at temperatures between about 95 and 115 degrees—or about the temperature of a hot tub, or a water-heater that's running a tad cool. When this warm water sprays or splashes, it launches little droplets carrying *Legionella* into the air. *Legionella* is so common in nature that most of us have inhaled our fair share. Fortunately, only a small percentage of us react with fever, chills, and cough. The victims tend to be over age fifty and are often smokers or heavy drinkers. Among this crowd the dust disease is a moderate killer, with a 5 to 25 percent fatality rate. Curiously, the same airborne bug can also cause a much quicker, less ageist, and less deadly illness called Pontiac fever.

The humidifier is another appliance that may spit troubling, water-wrapped dusts. This device *intentionally* throws tiny drops of water into the air of dry houses. Under normal circumstances these microscopic water beads carry only the minerals and metals that normally swim in tap water. As each water ball evaporates, grains of mineral dust are left hanging in the air. In areas with mineral-rich tap water the dust from a humidifier can easily exceed federal dust standards for outdoor air. This stuff settles as a faint white powder, and it reminds us that house dust can grow thicker even on a diet of water.

But on a *bad* humidifier day the drifting water balls will carry minerals *and* infectious little life-forms. The exact cause of "humidifier fever" isn't yet known. Investigators who track outbreaks of this disease often find humidifiers that harbor a confusing variety of organisms, from amoebae to bacteria to molds. The fever itself, which resembles a quick hit of the flu, is similar to diseases caused by endotoxins, the chemicals released by some bacteria. If the cause of humidifier fever is unclear, the federal government's advice to humidifier owners is not: clean and disinfect a humidifier tank as though it were the devil's personal hot tub.

And speaking of hot tubs, even *that* symbol of wealth and modern convenience is capable of burping up unpleasant dust. At an international medical society meeting in the spring of 2000, lung specialist Cecile Rose, a professor at the University of Colorado, accused hot tubs of spreading a tuberculosislike bacterial disease. The dust in question, she said, is a "nontubercular mycobacterium," normally a tidal-pool bug. When the bacterium multiplies in a hot tub, bursting bubbles can fling the cells into the air. When hot-tubbers inhale enough of these living dusts, they'll soon suffer fever, fatigue, night sweats, coughing, and even weight loss. Of the nine cases Rose had seen, she reported that four people had to be hospitalized.

With wealth itself spreading like a virus, Rose warned that her fellow docs would start seeing more of the hot-tub disease. Her advice to those stricken with what might be called "hot-tub lung" was to turn back the march of progress: take the tub outside. Outdoors, the natural air currents will keep the dangerous droplets from swirling too thickly. If the tub must stay in, Rose counseled, at least keep it tightly covered.

But the wheezing disease? Modern appliances can certainly contribute disease-causing dusts to the household. But researchers haven't linked these dusts with the proliferation of asthma.

Thus far our sifting through house dust has isolated some dusts that can cause cancer and various lung diseases. But we have not turned up a smoking gun in the asthma case. However, we have yet to consider a more ferocious element of modern dust. No, it's not the creepy dust mites or the tiny dust-dwelling predators that stalk them. Modern house dust is powdered with toxic chemicals. They account for a very small percentage of a typical house-dust sample. But they're potent.

"I never, ever used to think about dust—*never!*" says Paul Lioy, a high-speed man with an energetic head of black-and-gray curls. "Dust was this thing you put a broom down and swept up. We were just fortunate to discover that it's a really neat indicator of a house's history."

Lioy is an environmental scientist with both Rutgers University and the New Jersey–based Environmental and Occupational Health Sciences Institute. He and his colleagues discovered dust's tattletale quality when they measured the amount of toxic chromium in people's house dust during the cleanup of a nearby chromium-waste site. Cleaning up the outdoor environment, they discovered, eliminated much of the chromium from the house dust over the course of a year.

Although some toxic dusts sneak into our homes uninvited, Lioy is quick to note that others—indeed most—we tote inside ourselves. Since we're all partial to different poisons, and since our neighborhood dusts vary, no two homes host quite the same dust. But every one has some dark secrets.

About a decade ago the EPA announced the surprising news that the air inside most homes is dirtier than the air outside—a hundred times dirtier in some houses. And dirty air makes for richer dust. The EPA ranks our polluted homes among the top five environmental threats to human health. The agency blames tighter houses and toxic chemicals that are stealing in like an army of Greeks hidden in a herd of Trojan horses.

Formaldehyde arrives in the particle-board frame of a new sofa. Perchloroethylene enters in dry-cleaned clothes. In a particularly subtle invasion, lead has sometimes been known to sneak into houses in plastic miniblinds that silently shed toxic dust as they age. Para-dichlorobenzene turns up in the form of mothballs, and a platoon of pesticides creep in disguised as flea powder, roach bait, termite killer, and other modern conveniences.

Once these chemicals are inside, our houses act like pollution preserves. Some of the poisons, if they were left out in the elements, would break down. But the indoor world is a safe haven, where a zoo of chemicals take shelter from nature's hostility.

Pesticides account for a big percentage of the toxic dusts we bring indoors, says Lioy. More than 90 percent of U.S. homes harbor at least one flavor, be it flea powder, mosquito spray, termite killer, rosebush powder, or disinfectant. These home helpers may account for 80 percent of our *total* exposure to pesticides.

Lioy and colleagues have demonstrated how a pesticide dust can haunt a house. In 1998 the investigators reported that they had treated two apartments with a popular pesticide, chlorpyrifos. They studiously followed the instructions on the label, opening windows and running fans for four hours after spreading the chemical. For good measure they gave the treated rooms an extra hour to air. Then they set some plastic toys and stuffed animals in the rooms.

Each hour they pulled out one plastic and one cloth toy and measured the amount of pesticide that had accumulated on each toy's surface. Rather than fading with time, the pesticide clung more thickly with every passing hour, especially to the plush toys. A day and a half after the treatment, the rate of toxic dustfall finally hit its peak. Smaller amounts were still settling on surfaces two weeks after the treatment. This was a shifty dust.

"Pesticides are very dynamic chemicals," Lioy concludes. "They *move*."

Not every toxic dust waits for an invitation to enter the pollution preserve. If the air in your neighborhood features toxic chemicals, they'll blow in when you open the door. Soil tainted with poisons will hitch a ride in on your feet. And that helps to explain why pesticides that were banned decades ago can still be found in the dust of many houses.

DDT, for instance, was a popular pesticide in the United States until it was shown to be accumulating in the bodies of bald eagles and other raptors. The pesticide had saturated the food chain to the point that the poisoned birds were unable to produce crushproof eggshells. In 1972 DDT was banned in the United States. Twenty years later researchers went looking for the infamous pesticide in the carpets of a few hundred midwestern houses. When the researchers vacuumed with special equipment, one in four homes coughed up dust spiked with DDT.

PCBs, another group of industrial poisons, are similarly persistent. Although PCBs are no longer manufactured in the United States, studies suggest that tiny amounts of them still linger in the air and dust of every house in the nation.

Likewise lead. Some of the lead in house dust comes from old paint inside the house, but Lioy says that even more may come from outdoors. Just walking across the face of the Earth, he says, you gather dirt and dust on your shoes. If you live near a road where dusts of the leaded-gas era linger, or if there's a lead-related industry nearby, the dust on your shoes will be lead-rich. Each time you cross the threshold, you'll carry in more. As you walk across the floor, microscopic earthquakes and avalanches of leaded dust will tumble off your shoes. Lead also *flies* in, judging from a study of attic dust in Nevada and Utah. This study found that the greater the age of the house, the more lead dust there is in the attic. And the lead in the attics was chemically distinct from the lead in the nearby soil, which suggested the attics were gathering well-traveled airborne dust.

Along with the rare old species, modern chemicals barge in, too. Toxic chromium and mercury metals are still widely produced, and they are a common component of house dust. In many cases scientists find that these metals are much more concentrated indoors than they are in the dust sampled outside the house. Lawn pesticides and herbicides also make a strong showing.

Plain old dirt dust is, of course, the number-one import. It generally accounts for about half of the total amount of dust indoors. But even that can be strangely toxic. The same research team that has identified mutagenic compounds in cooking meat accidentally stumbled across mutagenic *dirt*. The researchers assumed that someone had spilled laboratory materials outdoors. But to be sure, they gathered more soil samples, far from the lab. "My backyard, someone else's backyard," marvels Lawrence Livermore biologist James Felton, "*all* the samples showed mutagenicity!" Felton speculates that the mutagenic compounds may be made by soil organisms, or they could be agricultural chemicals gone AWOL.

Of all the toxic chemicals in house dust, however, none is so enthusiastically produced as tobacco smoke. An estimated 48 million adults in the United States smoke. In 1999 these people reduced 435 billion cigarettes to smoke and ash. Nearly 4 billion cigars, as well as untold pipeloads of tobacco, added their dust to the national smoke cloud. Most of these little fires were kindled indoors. And this is rich dust. It's not hyperbole—there really *are* 4,000 chemicals in tobacco smoke. Fifty of them are known to cause cancer. And this toxic dust is easily detected in the house dust of smokers. Every house has a background level of tiny dusts in the air. In houses that shelter a smoker, that background level is about doubled.

Much of this dust settles in people's lungs. The black specks stick to the lining of the bronchial plumbing and to the air sacs, or alveoli. They darken the lungs. And they sabotage the mucous escalator that ships dust out of the lungs. The dust that has collected on its sticky surface lingers, leaking compounds into the body.

No dust can compete with tobacco smoke for deadliness: this dust kills almost half a million smokers in the United States every year. Secondhand smoke kills an additional 3,000 nonsmokers. And each year these dusts may cause pneumonia or bronchitis in 300,000 infants. The smoke even fouls outdoor air, with health consequences that aren't known: a 1994 report estimated that in the category of tiniest particles, cigarette smoke accounted for about one in a hundred pieces of dust in the outdoor air of Los Angeles.

And then there is asthma.

Tobacco smoke is thought to aggravate the preexisting asthma of about a million children. But more damning is the fact that a child in a smoky home is likelier to *become* asthmatic than a child in a smoke-free home. To a scientist such an association does not amount to a clear case of cause and effect. And even if tobacco smoke were proven to *cause* asthma, the dust of burned tobacco alone couldn't explain the entire asthma blaze. Smoking in the United States, after all, has been on the wane, while asthma has been on the rise. Tobacco smoke may contribute to the asthma disaster. But it couldn't possibly be the solitary culprit.

As all these dangerous dusts settle, they accumulate on the floor—at baby level, in other words. Especially in carpets, even in well-kept houses, toxic dust can become dangerously concentrated. Dust that settles on a bare floor is vulnerable to vacuuming and other harassments. But dust that lands in a carpet can burrow into the forest of fibers until no casual cleaner can dislodge it.

John Roberts is a Seattle dust expert and consultant to a Seattle program that attacks dust in the homes of low-income asthmatic children. And he knows that you have a filthy carpet.

"Normal vacuuming does not remove deep dust," he says flatly, impatiently. "In fact, it allows deep dust to build up to deep, *deep* levels. In a study of older carpets we have found between eight and one hundred seventy grams of deep dust per square meter of carpet." That's between three tablespoons and three and a half cups of dust. This deep well of old dust, Roberts says, can rise up under foot traffic to foul the air anew. And a quick vacuuming job can bring it to the surface—and leave it there, where crawling children can gather it.

"The risks are not small for a crawling child," he insists. "For example, the best predictor of how much lead a child has in his blood is how much lead is in the carpet dust. Furthermore, we calculate that the average infant gets the same amount of benzo-a-pyrene—a strong carcinogen—from dust that he would get from smoking three cigarettes a day. And that's the *average* carpet. Some carpets have *one hundred times* that amount."

And kids do consume this dust, hand over fist. Based on current estimates, by the time a child is six, he or she may have eaten half a cup or more of fine dust—that's dust minus the big stuff like dog hair, sand, toast crumbs, and sweater lint. This will not surprise most parents. Crawling-age kids seem to spend half their time chugging around on their sticky little paws and the other half remoistening those paws by sucking on them. *Of course* babies eat dust and dirt. Most parents accept this as an immutable and unalarming law of childhood.

But our dust is not the dust of a few decades ago. Sure, previous generations of babies cut their teeth on a bit of lead dust and tobacco smoke, not to mention horse manure blown in from unpaved streets. But our tighter houses and broader swaths of dust-rich carpeting, combined with our wholehearted fling with chemicals, concentrate poisonous dusts right where children are most like to slap a sticky hand on them. The upshot is that for children under six, house dust accounts for a huge portion of their total exposure to toxic chemicals.

Natalie Freeman, a psychobiologist at the Robert Wood Johnson Medical School in New Jersey, studies the fine points of children's relationship with dust. One of her recent experiments measured the rate at which two-year-olds transfer dust from their hands to their food. On the first day of the experiment investigators wiped the children's hands clean and measured the amount of dust they captured.

"Usually less than ten milligrams," Freeman reports. "A few very grubby kids had about sixty milligrams." (Ten milligrams is one or two hundredths of a teaspoon.) On the second day of the experiment, when the kids had naturally resurfaced their hands with house dust, they were instructed to take a hot dog or a banana from a plastic bag, break it into pieces, and put the pieces back into the bag.

"The lead on the food correlated very nicely with the lead we got from the hand wipes," Freeman concludes soberly. And that was without the "eagle claw" or the "fist."

"Kids under two years old use what I call 'eagle claws' to grasp the food; then they eat what sticks out of the fingers," Freeman says. "Two-year-olds wrap food in their fist, with complete palm and finger contact." Both methods would probably transfer more dust from the hand to the food. But when the kids broke apart the test foods, they used mainly their fingertips, which probably falsely lowered the dust transfer. Nor did they drop the food on the floor, which Freeman says is another popular method for conveying dust to the mouth.

Even without food as an intermediary, kids eat a lot of dust. Videotaping experiments have shown that kids between two and five years old put a hand in their mouth an average of almost ten times an hour. Superslurpy kids double that rate. Based on this "hand-to-mouth activity," investigators estimate that the average child eats 15 or 20 milligrams of dust a day, and that the super-slurpers eat 30 to 50 milligrams.

"It may not sound like much," says Freeman. "But over the years it adds up." What does it add up *to*? Given the regrettable absence of scientific data on this subject, a reckless journalist might resort to guerrilla math: the average kid, who ingests 15 milligrams of dust a day from his first birthday to his sixth, might consume about half a cup of fine, dense dust. Over the same time the slurpiest kid will gulp down one and two-thirds cups of dust.

Apparently, scientific proposals to estimate a person's *lifetime* dust consumption have also lost the battle for research funding. But if we assume (as some researchers do and some do not) that older children and adults eat about one-fifth the dust that younger children devour, then the average seventy-six-year-old will have swallowed about two and a half cups of dust. And the seventy-six-year-old who began life with extra hand-to-mouth activity will have swallowed three and a half cups.

All this might be more comical if it weren't for the perilous contaminants in house dust. Lead, pesticides, mutagenic soots, PCBs, the four thousand products of tobacco smoke—in the proper dose some of these chemicals can cause everything from mental retardation to nerve damage to cancer and lung diseases. But curiously, beyond the asthma-inspiring qualities of burned to-bacco, there are no other smoking guns in this chemical menagerie.

But a flock of clues to the asthma mystery may turn up in the final category of house dusts: the dusts of living things.

Molds are like grasses that sprout from the rich soil of our dust. Just as molds break down dead leaves outdoors, they also clean up organic messes indoors. Nearly half of the average dust sample is made up of items a fungus might favor. That includes a tangle of fabric fibers—tiny multicolored sheddings from sweaters and sheets, pillows and sofa cushions, towels and rugs. The millions of skin particles that you and your pets shed every hour make a generous contri-bution. And skin flakes are roughly matched by minute cellulose fibers, includ-ing everything from tea leaves to onion skins, from shreds of toilet paper to specks of the *Times*. Furthermore, fungi can cultivate the paper coating on dry-wall, the soap film on a shower curtain, damp towels, the carpet, the mattress. Some of the gases that molds emit give damp rooms their musty smell. Al-though most species require a dab of moisture, most houses have enough.

When they're thriving, molds launch a fine dust of spores into the air, with the goal of enlarging their empire. And increasingly, these molds and the thou-sands of chemicals they release as they grow are blamed for sickening both buildings and people.

Some mold by-products—aflatoxins, for example—are among the most powerful carcinogens invented by man or nature. Other mold chemicals are toxic to lung tissues, if they invade the body in adequate quantity. All fungal spores contain proteins—and proteins can cause allergies. The mean side of in-door fungi has only recently come to light. But they're now under heavy suspi-cion. Health investigators used to search "sick buildings" for noxious carpet glues and toxic cleaning products. Now when schoolchildren sniffle or desk jockeys stampede for home with headaches, one of the first questions the health investigators ask is, Where's the mold?

In the average house the amount of mold drifting in the air can be consider-able. Experiments that count the number of mold spores in houses find that 1,000 "colony-forming units" per cubic meter of air is quite normal. (For prac-tical purposes a "colony-forming unit" is a healthy fungal spore.) A Kansas City survey found that half the homes tested had ten times that much mold in their air. And truly sky-high mold counts are common in houses with damp basements or water damage.

Some molds seem able to cause symptoms—itchy eyes perhaps, or a sore throat—at the rather low level of 100 spores per cubic meter of air. Others may need to swarm in clouds of 3,000 per cubic meter to induce the same amount of irritation.

And lately, mold dusts have been caught causing symptoms that may relate to asthma. "Chronic rhinosinusitis" (CRS) has been a puzzling condition. The number of people suffering from perennially stuffy noses and sinuses grew 60 percent between 1982 and 1994. That's the same terrific growth rate as asthma. CRS now afflicts roughly 40 million people in the United States.

Jens Ponikau is an ear-nose-throat researcher at the Mayo Clinic in Rochester, Minnesota. He and colleagues recently discovered that mold plays a role in CRS. But, more intriguing, they concluded that it is the body's unnecessarily violent *response* to mold that causes the problem.

We all have fungi in our noses, after all—at least two different species on average. Why do some people's immune systems get hysterical about these dusts? In the nose of a person with CRS, these fungi inspire an attack response in an immune-system cell called an eosinophil. "You can see it," Ponikau says in a crisp German accent. "We can demonstrate that the eosinophils are actually going out to attack the fungi. These eosinophils release toxins, which erode the nasal membrane. And then bacteria can invade. It's just like when you make a cut in your hand: the bacteria can infect it."

Until recently doctors couldn't do much about CRS. They often tried antibiotics, Ponikau says, on the theory that bacteria cause the problem. But relief was temporary.

"Two weeks later they're back: 'Doctor, I have a sinus infection,'" Ponikau recounts. Now Ponikau's team is using antifungal nose spray to keep CRS at bay. But this is not a cure. And just like asthma, CRS continues to explode.

Ponikau has a theory, and he says it's related to the asthma epidemic as well. But for the time being he's keeping this moldy little secret up his sleeve. To let a theory out before he can furnish some solid evidence would be an act of professional suicide.

"It's hot stuff, and we want to be sure we're on the right track," he says. "Believe me, when we release this, there will be resistance from industry—and the public. I can only tell you it's something we *do*," he hints with a chuckle.

If Ponikau is right, then something about life in modernized nations, and especially in cities, is making us react strangely to our house dust. Something we do is sabotaging our ability to shake off some of the most common dusts in our homes.

. . .

If the thought of mold webbing its way across your floor, through your bed, and into your nose is unsettling, there is a bright side: a variety of little creatures are eating it. If mold is grass, then cows are plentiful. Most plentiful are the dust mites.

Almost everyone has seen a picture of a dust mite by now. They're shaped like softening balloons. They stand on arched, pointy legs. And instead of culminating in a proper head, their bodies are led by a collection of fingerlike feeding tools. In the pictures they're always gray. But alive and in person they're much more attractive. They're cream-colored and glossy. They scamper over carpet fibers with a surprisingly industrious spirit. And well they should. Cleaning up all the skin you shed is no small chore.

In temperate climates two species of mite devote themselves mainly to skin eating, although they probably eat fungi and other tidbits they stumble across as well. To most people the two are indistinguishable. But Larry Arlian is not like most people. Outside his lab at Wright State University in Dayton, Ohio, a nameplate reads MITES R US. The man himself is a bit on the quiet side.

"I suppose we have many millions of mites here, if you want to count individuals," he says. "Well, actually, there are probably a million in a *single* culture. In a thriving culture you can have several inches of mites and medium [mite food]. They don't seem to mind walking on each other."

Arlian rears the beasts both to study the powerful allergens they produce and to learn more about their biology. Dust mites are large, by dust standards. They're about three hair-widths long, and only a few dozen of them could crowd onto the head of a pin. You could probably watch them graze in your bed, if you had supersmooth black sheets.

"If you release them on a black table, you can see them move around," Arlian says. "If you hold a glass vial up to the light, you can see them." But your odds of spotting these creatures in the wild—in your bed, couch, or carpet—are long. Not only do they shun open terrain, but they also seem to take very few breaks from their life's work, which is eating.

"They're not parasites," emphasizes the man who has been known to transfer a few of the creamy little wanderers to the back of his hand so that guests can admire them under the microscope. There they tirelessly clamber up one hair and down the next, in a fruitless search for food.

Females may eat half their body weight a day. For a human adult that would be the loose equivalent of seventy-five pounds of skin. The mite spits a softening substance onto a skin flake—which might be larger or smaller than she is—and then blindly claws the liquefied food into her mouth. Thus fueled, females lay an average of two or three pale, sticky eggs a day.

And all this bustle takes place in dusty bookshelves, in the carpet, and in your pillow as you sleep at night. Dust mites can make a living almost anywhere in the house. But their numbers are thickest where the skin is richest. In such places a teaspoon of fine dust might hold between five hundred and a thousand mites.

The bed, surprisingly, is seldom the most attractive habitat. Mites do crawl around in mattresses, sheets, and pillows. But on one of Arlian's mite-hunting safaris he found that the bed was the biggest attraction in just one-fifth of the homes he investigated. Furthermore, the mites don't seem to enjoy close personal contact: they stay on the opposite side of the sheet from you or restrict their travel to the inside of the mattress or pillow. Many are the rumors that elderly pillows are virtually stuffed with dead dust mites. Ten percent is a common statistic, although some alarmists claim that fully 25 percent of an old pillow is old mites. "Nonsense," says Arlian to both guesstimates. "I don't think anyone really has done the analysis. But my sense is that it's a *very* small amount."

A house's highest mite counts are typically found between couch cushions, on the family-room floor, and on the bedroom floor. On rare occasions Arlian will discover a house where a dust sample from one of these spots produces about *18,000* mites per teaspoon of dust.

There is a strong relationship between asthma and the dust of these little scavengers. As with tobacco smoke, there's a chance that an overload of dust-mite dust may *cause* asthma in children. It clearly causes allergies.

About 15 or 20 million people in the United States are allergic to dust mites—or, as allergists delicately phrase it, to "dust-mite allergen." The mites themselves are too big to sail through the air and up the nose. But their manure and their decomposing body parts are not. The manure exits the mite as brown balls one-sixth of a hair's width in diameter. The balls consist of digestive enzymes plus digested remnants of all the things a mite might eat. One investigation of the contents of dust-mite tummies found that they seem to swallow everything from pollen to fungi, bacteria, plant fibers, moth and butterfly scales, bird-skin scales, and yeast cells.

About twenty times a day a mite excretes a package holding three to five of these balls. Proteins in these packages are the source of much human misery. Additional allergens haunt the broken bits of dead mites, too. And unlike the mites themselves, these small items rise easily into the air. They dance in the sunlight when you make the bed. They rise in a cloud when you plop down onto the couch. Once stirred up, they'll float for twenty or thirty minutes.

Even when the individual dung balls break free of their package, they're still too big to enter the lungs. But they, and other mite parts, stick to the moist lining of the nose, where they cause abundant havoc. As with any allergy, it is the body's unnecessarily violent response to these intruding dusts that causes the suffering.

Some studies suggest that children who live in extra-dust-mitey homes are likelier to develop asthma. Other studies negate this. But even if dust-mite dust should prove to trigger the development of asthma, that couldn't account for all the new cases.

The Achilles' heel of a dust mite is its need for water. When mites don't get enough from your skin scales, they have to trap their water, one molecule at a time. From a gland near its mouth a mite excretes a salty solution that pulls moisture from the air. This brine trickles into a mite's maw. But much of the western United States is naturally too arid for a mite to gather water—yet asthma is thriving there. And mites also shun cold, dry Finland, where asthma is raging.

Many other house-dust inhabitants may also produce allergens. Johanna E. M. H. van Bronswijk is a professor of public-health engineering at Eindhoven University of Technology, in southern Holland. In her wonderfully droll textbook, *House Dust Biology,* van Bronswijk devotes an entire chapter to what she calls "the house dust ecosystem." It adds up to a who-eats-whom guide to house dust.

The fungi fill the biological role of "decomposers," she notes, disposing of everything from skin to deceased dust creatures. "Furniture mites" likewise specialize in decomposing organic matter, including cotton stuffing, wood flooring, paper, reed mats, and the like.

The fungi are mowed down by a variety of creatures, including dust mites and book lice. (Book lice are tiny, speedy insects, often seen zipping across old papers.) And the mites and book lice are hunted by a few different predators.

One of those, *Cheyletus eruditus,* was first discovered in a library, and hence its bookish moniker. But this mite, whose Latin name translates to "crab-clawed, erudite," is no fainting violet. To the untrained eye, *C. eruditus* looks like a typical balloon-shaped and headless mite. But it's bigger than a dust mite, and a professional can easily see other differences.

"They are fun," van Bronswijk writes in an e-mail note. "They hide under particles, with only their massive, scissor-like tools visible. When hungry, they grab a passing mite with their scissors. Then they introduce a stiletto-like tool in the prey, and apply lytic enzymes that liquefy the content of the inner mite body. Subsequently, they suck the victim empty, and only the dead skin remains." In addition to mites, these carnivores eat book lice and flea larvae, and when the chips are down, they eat one another.

Larry Arlian, the dust-mite man, finds predatory mites such ferocious hunters that he's forced to position his dust-mite cages inside moats of oil and to grease the sides of the cages with petroleum jelly. Regardless, these bloodthirsty animals will sometimes break into a cage and begin devouring the experiments. And like the dust mites they kill, these predatory mites may also produce allergenic dusts.

"Pseudoscorpions" are equally savage hunters. And they look the part. They're crab-shaped, armed with long claws, and visible to the naked eye. The book scorpion is one of this clan. In a dark crevice, such as the black canyon between books, it awaits the innocent dust mite or book louse that might pass by. And then it pounces, claws slashing.

Moving up the dust food chain, van Bronswijk presents fleas. In houses with cats or dogs, flea larvae may squirm through the dust, selecting and swallowing the feces of their elders. They are also reputed to scavenge dead dust mites—and in hard times, each other. The presence of furry pets also assures that house dust will be enriched with animal dander—yet another allergen that is at least *associated* with asthma, if not proven to cause it.

Next come silverfish. These are suspected of dining on the occasional dust mite, but analysis of the gut of a silverfish suggests that these critters will swallow almost any sort of dust. Inside one silverfish, van Bronswijk recounts in *House Dust Biology,* were found "fragments of plant tissue, sand grains, pollen grains, *Protococcus* [bacteria], fungal spores . . . and hyphae, starch grains, animals hairs, setae, scales and tracheae of arthropods." Silverfish are also capable of digesting cotton fiber, paper products, and rayon.

Near the top of the dustheap are cockroaches. These creatures also produce tiny fecal dusts that cause allergies. But once again science is unclear on whether a child who breathes an excess of cockroach dusts stands a higher risk of developing asthma, too.

This diverse community of living things inhabits dust all over the house. And the bed, according to van Bronswijk, provides a special microhabitat. The menu of straight skin flakes, fungi, and cotton fibers may be monotonous. But as the floors of modern homes become drier, beds could offer a humid refuge for fungi, mites, and other small and sensitive forms of life. Van Bronswijk reports that one investigator has even grown a fern from a sample of bed dust—in her book she includes a photograph of the specimen, nicely potted.

. . .

House dust holds a bit of everything under the Sun. It is animal, it is vegetable, it is mineral. It is a durable catalog of modern chemicals. But what is its role in the asthma epidemic? Although a dust ball is riddled with hints and possibilities, researchers have yet to find one ingredient that can explain all the wheezing.

On one hand, it's clear that people in wealthy nations are living in tighter houses that concentrate all the elements of house dust, including common allergens and industrial-age chemicals. Overexposure to certain proteins can cause allergies. And allergies are a risk factor for developing asthma.

On the other hand, no single item in house dust can bear the blame. And so the most promising theories about the asthma boom argue that what has changed is the way our modern bodies react to dust.

To examine one of these theories, let's return to Virginia, where Thomas Platts-Mills contemplates the distant blue mountains through his office window. Platts-Mills thinks the difference is that our underused lungs are losing their edge.

"In 1950 kids spent twenty hours a day indoors," he says. "In 1990 they spent twenty-three and a half hours indoors. It's a small difference, percentage-wise, right?" It is a modest change: kids increased their indoor time by less than 25 percent.

Platts-Mills taps his desk for emphasis.

"But now, four hours outside, versus a half hour outside? That's the *big* change." That's nearly 90 percent less time spent outdoors. What Platts-Mills is proposing is that being indoors isn't nearly as bad for children as *not* being *outdoors*. And the primary benefit of being outdoors, he says, is that kids move.

"We've discovered a way to make children sit still," he says, refocusing his sharp gaze. "Visual entertainment. Exercise is an anti-inflammatory. And I suspect that it accelerates healing."

By exercise Platts-Mills does not mean mountain climbing and jogging. He means "gradual or repeated activity"—in a word, playing. Even walking would qualify. "The data show that in villages where walking remains the normal form of transport, asthma is still rare," Platts-Mills wrote in a 1998 research paper. "In Papua, New Guinea, rural Africa, Aboriginal Australia, and among Inuit Eskimos, asthma is uncommon."

By contrast, he suggests, sitting still in front of the television may allow an unprecedented degree of sluggishness to settle into the body. It turns out that taking the occasional deep breath is a gentle form of exercise for the lungs. But people take fewer of those when they sit reading, Platts-Mills says, and fewer still when they watch television.

"They sigh slightly less often than if they were reading a book. It's not *absolutely* clear that that's a disaster," he says dubiously. But it's a hint. A stronger condemnation of the chatterbox is that it's associated with obesity, Platts-Mills adds. And he says a scientific correlation between obesity and asthma is beginning to take shape.

So perhaps sitting still is the accomplice that all the lung-teasing elements of house dust have in common. Perhaps all these dusts need is the kind of lazy lung that only a sophisticated and sedentary society can produce.

That's one theory.

In a poem published in 1912, Ezra Pound may have summarized another theory, long before it was formally named. Pound contrasted an upper-class Londoner with what he called "the filthy, sturdy, unkillable infants of the very poor.

"They shall inherit the earth," he mused.

Pound is infamous for his fascist sympathies, but he did have a way with words. Friedrich Nietzsche had earlier voiced a similar, if less poetic, observation: "What doesn't kill me makes me stronger."

And that is the main point of the "hygiene hypothesis." This proposal suggests that life in the industrialized nations is not dusty *enough*. In spiffed-up spots all over the world, this regrettable lack of grime is producing crowds of weaklings.

More precisely, a fast-growing body of research suggests that early bouts with some germs and parasites can harden a child's impressionable immune system against the development of allergies and asthma. But as a developing nation cleans up, its children endure fewer diseases, and their immune systems fail to mature. Then a little bit of dust can push their bodies into a hysterical reaction.

It is an extremely hot theory. Each month medical journals deliver fresh studies of new groups of children. Most of these revelations are controversial, to some degree. They include:

- Various groups of European farm children, who presumably spend more quality time with mold, manure, and other rough dusts, suffer less allergy and asthma than their nonfarming school chums.

- Italian air force cadets whose blood reveals evidence of old episodes of

food poisoning and stomach bugs are less likely than their bugless bud-
dies to have asthma.

- Denver-area children from homes with high levels of endotoxin—bacterial poisons—are less likely to develop allergies and, presumably, asthma.

- Children who get measles before they turn three are less likely to develop
  asthma. Similar connections have been made for infections with hepatitis
  A, tuberculosis, influenza, moldlike "mycobacteria," and even some vacci-
  nations.

- An English study reported that children who take antibiotics in the first
  two years of life are at higher risk for asthma. The authors speculated that
  antibiotics may indiscriminately kill off bacteria that toughen the immune system.

- Even parasites may steel the immune system against allergenic dusts. A
  study of poor children in Venezuela found that the children with the
  most intestinal worms were least likely to have dust-mite allergies. After a
  group of the children took deworming drugs, the rate of dust-mite allergy
  quadrupled in that group.

What have all these diseases to do with hygiene? The hygiene hypothesis
sprouted from the observation that families are getting smaller, and therefore
siblings aren't swapping as many germs. Old-fashioned crowding gave children more chances to inhale each other's respiratory germs and to trade what
researchers call "orofecal microbes," which produce gut diseases. In the summer of 2000 this observation was bolstered by a big study of Arizona children.
Those children with at least one older sibling, or who went to day care before
they were six months old, halved their likelihood of being asthmatic at age
thirteen.

The hygiene hypothesis has broadened to embrace a variety of modern practices. The researchers who studied the Italian cadets, for instance, pointed to
"a Westernized, semisterile diet" that presents no challenge to a child's developing immune system.

Poverty is also an element in the hypothesis. When the Berlin Wall fell, for
instance, health investigators discovered that asthma was common in wealthy
West Germany but rare in polluted and impoverished East Germany. Asth-
matically speaking, East Germans are now catching up with their wealthy
brethren. Similarly, a study in Ethiopia found that people who lived traditional, earthy lives in small villages displayed a higher rate of dust-mite allergy

than people living in the city of Jimma—but the villagers suffered none of the *asthma* that also plagued the city folk.

So by bits and pieces researchers are building a case for *how* Ezra Pound's very poor and filthy children might have earned their unkillability. Kids who live dust-rich lives may have more balanced immune systems.

"It's intriguing to think that there are some things in dust that can educate the immune system," says Andrew Liu, a physician and asthma specialist at the National Jewish Medical and Research Center in Denver, Colorado. Liu, who has a wide smile and a shock of black hair, is an author of the study that found that kids in Denver homes with lots of bacterial endotoxins may be at a lower risk for developing asthma. So he's partial to the notion that endotoxins, poisons from bacteria like salmonella and *E. coli,* will eventually prove to be a powerful controlling force on asthma.

"You don't need to be infected by germs to stimulate the immune system," he proposes. All a child needs to do is cuddle up to a dust bunny that's rich in endotoxin dusts. In fact, Liu can imagine a day when those babies who don't have enough dust at home might be inoculated with endotoxin to fend off asthma—though he confesses with a chuckle that he'd be "too chicken" to give the first shot. Other researchers have suggested that inoculations with mycobacteria—of which tuberculosis is one—might also be used to prevent asthma.

It may sound bizarre, but that wheel was accidentally invented decades ago. Agile Redmon is a semiretired Texas allergist. He recalls the early days of treating "house-dust allergy," before doctors knew exactly which elements of dust their patients were reacting to. "Sometimes we'd take dust from a patient's vacuum-cleaner bag," he says. "We'd sterilize it, grind it up, and make a solution. Then we'd use this extract in immunotherapy. I'm sure we were giving endotoxin in our house dust," Redmon muses. "It worked."

People are still shot up with house dust, in fact. And at least one manufacturer of house-dust extract still gets its dust from vacuum-cleaner bags: churches and schools that surround Greer Laboratories in Lenoir, North Carolina, collect the dust bags from the public and sell them to the company at two dollars per pound of dust.

So bacterial poisons in our dust may prevent asthma. It's a theory.

"The whole history of medicine is about observing the epidemiology of a disease and then coming up with a whole *lot* of theories that might explain it," says Andy Liu. And then people try to tease out the really important details.

"Prove me wrong," he invites cheerfully. "That's what science is about. Go ahead! Prove me wrong!"

· · ·

Proving that *any* of these theories is wrong will be a trick. Consider one snapshot of the dust-and-asthma puzzle—and observe how the competing theories chase each other around it.

Among New York City's homeless children, more than one in three kids have asthma. At some schools in the Bronx the rate is closer to one child in six: still about three times the national average. As of the mid-1990s about eleven New York City children died of asthma in an average year. So these days not every poor child is unkillable. And what light can the competing theories shed upon this microcosm of the asthma epidemic?

The quaint old notion of tight houses and concentrated dust actually fits here. There is solid evidence that inner-city housing is often extra thick with mites, cockroaches, mice, rats, or mold—all of which make allergenic dusts. And dust allergies are associated with asthma.

But the "sitting still" hypothesis makes an equally good case: Thomas Platts-Mills says that kids in crime-ridden neighborhoods are exercising even less than the rest of this sedentary nation.

"We're the only nation in the world where activity is lower in the lower class than in the other classes," he says. "In 1970 there was no class difference in the asthma mortality rate. In 1990 there's a difference. But *only* in the U.S."

And the hygiene hypothesis makes a strong showing, too. Suppose that New York's poorest children are the *least* likely to encounter the endotoxin dusts that farm children play in. Suppose they *do* eat a "sterile, Western diet." In the United States at least, poverty alone can't guarantee grubbiness.

There simply is no obvious explanation. For now the answer to how house dust relates to asthma is a tantalizing secret. For now the theories of dust and disease dance in the air like the specks that flow through a house when dinner's in the oven, the candles are flickering, and the pseudoscorpions are seeking out their evening meal in the deep forest of the living-room carpet.

11

DUST TO DUST

A human body is mainly water and bone. Bone is mainly calcium phosphate, plus traces of other elements, including stored pollutants like lead. The watery parts are tinged with carbon and nitrogen, iron and sulfur, chlorine and sodium, and a suite of trace elements from arsenic to zinc. All of these elements, of course, originated in space and were bundled into the planet during the birth of the solar system. They're yours for as long as you live.

But as soon as you die, your borrowed elements start to slip back out of your body, to recirculate. Even people who go in for modern mummification and storage in a stainless-steel pod aren't going to last forever. When the Sun begins to throb like an overtaxed heart, there will be no exceptions to the rule: Dust thou art, and unto dust shalt thou return.

At room temperature, microbes quickly begin breaking apart the cells of a dead body. In fact, those cells can fall apart all by themselves. Decay releases fluids and gases. Fungi may quickly colonize dead skin, transforming flesh directly into a little cloud of fungal spores.

Most dead bodies are popped into a fridge at the morgue, where they will stay "fresh" for many days. But in the United States especially, many families wage impassioned combat against a loved one's impending state of dust. By pumping a body full of formaldehyde and other preservatives, an embalmer can kill off many of the destructive bugs and give dead tissue a lifelike rigidity. With a layer of makeup, plus various stuffings, glue, mouth formers, eyelid formers and other props, the body can retain a lifelike appearance for a few more days—for much longer in some cases.

Sealing an embalmed body in a casket can further postpone its disintegration. However, the traditional casket isn't terribly resistant to the underground forces of water and earth. So the dirt that lies atop a casket may quickly cave in the lid. That introduces a horde of bacteria, fungi, and other decomposers. This calamity has led to the widespread use of a concrete grave liner, which is essentially a casket for the casket. Thus sheltered, and in a dry climate, an embalmed body might escape disintegration for years.

But eventually, moisture will find a way into the casket. Little recyclers in the soil will set to work. And gradually, the elements of a body will sink into the surrounding soil, becoming part and parcel of the Earth's gritty skin.

The hard items in a casket resist the dust state the longest. The pieces of plastic and metal that both surgeons and embalmers tend to insert into modern bodies will break down slowly. The metal jewelry and zippers, the plastic buttons and shoes, and any durable mementos in the casket will also linger. And the bone. As the dinosaurs can attest, bone occasionally fights the dust-making forces for so long that trickling water has time to dissolve it very slowly. Bit by bit the water replaces soft bone with hard mineral.

Human bones are certainly eligible for fossilization. Should your body be buried quickly, and in exactly the right sort of soil, then in thousands to millions of years your skeleton could turn to stone. The fossils would not be your body, technically. The water would carry away your original bone molecules and mix them with the dirt and dust. The mineral replicas of your bones would be much more durable. However, even your mineralized fossil would crumble to dust when your grave was finally opened up by erosion. Fossil hunters typically make their discoveries when they stumble across a clutter of dustifying fossil on the Earth's surface. When they dig into the protective rock, they find the rest of the fossil intact.

If your bones don't win the fossil lottery, their crumbled and dusty end will probably arrive much sooner. Erosion will scrape open your former grave. Tiny bits of soil, perhaps stained dark with your iron and frosted with your calcium, will tumble downstream on a trickle of water, or they'll rise up and take to the wind. If the mechanics of the Earth's shifting plates send you down, not up, you'll be ground up and mixed with melting rock, perhaps to emerge later as volcanic ash.

Occasionally, carefully stored human bodies have been reduced to dust not by the restless Earth but by their fellow man. Europeans, from the Middle Ages through the eighteenth century, considered Egyptian mummies to be potent medicine, meaning that those ancients unlucky enough to be discovered were liable to be powdered and swallowed. Mummies have also been ground to dust for use as fertilizer, says Kenneth Iserson, author of *Death to Dust*, and they were once shipped by the boatload to the United States, where captains of industry experimented with shredding their cloth wrappings to make paper pulp. But: "The wrappings were too stained to produce a high quality of paper," says Iserson.

However, in spite of the efforts of mold, maggots, erosion, and even mummy grinders, burial is still the slow route to dust. If the grave is dry, and if erosion spares the landscape, burial can put a long stay of execution on the dust-to-dust order.

Excarnation, practiced in parts of India, Asia, and Africa, is faster. A dead body is usually positioned in a tree or at some other site dedicated to the purpose. Then animals, sometimes well accustomed to this ritual, move in to disassemble it. In Tibet, for instance, it's still common for a dead person's family to pay a team of undertakers to carry the body of a loved one to a hill and butcher it for waiting vultures.

Pamela Logan, a Southern Californian who raises money to aid Tibet, is one of the few outsiders who has witnessed what is locally known as a sky burial. "When the vultures saw the funeral party coming up the hill, they started circling," she recalls. On a stone platform, Logan says, the undertakers made some quick cuts in the body, then stood back. "About fifty enormous vultures moved in," she says. "In thirteen minutes there was no flesh left. Just gristle and bone. The men moved in with these big mallets and pounded the bones to a pulp." This the men mixed with flour and doled out to waiting crows and hawks when the gorged vultures had departed.

So in a matter of forty-five minutes, says Logan, the dead person's borrowed elements were incorporated into new bodies. For Tibetans who can't afford such a ceremony it is customary to leave a body in the hills for birds, dogs, bugs, and other creatures that might voluntarily perform the service.

As these winged, pawed, and pincered hosts digest the body, they absorb some elements and reject others. The rejected components exit the back end of the animal in a form that quickly dries to dust. A fraction of the absorbed elements, however, might evade dustification indefinitely. When the vulture host dies, a new circle of scavengers will take shares of its elements and carry them around for a while. And so some parts of a body will hop from vulture to dog, to fly larva before reaching dusthood—perhaps as a mold spore, launched from the platform of a dead fly.

Alas, due to a certain squeamishness, and concerns for public health, excarnation is not available to residents of the United States. For those who want to rush the body back to dust, cremation will have to do. If your paperwork is in order, you can be reduced to a wisp of smoke and few pounds of bone dust within a few hours of your death.

Burning is a time-honored method for processing a dead body. In ancient Greece the smoky practice was embraced because not only did it prevent disease, but it also prevented one's enemies from committing acts of disrespect against one's corpse. In Rome burning became so popular that the city fathers had to ban it from the city proper. The English word "bonfire" is left over from a time when the people of Britain burned their dead atop a "bone fire."

Regardless of the culture, the usual practice was to collect the pale bone fragments when the bone fire cooled. These remnants of a body might be buried or stored aboveground. In many cultures the bones and ashes weren't—and aren't—as important as the burning itself: the rising dusts and gases reinforce the common notion that destroying the body frees the soul.

But in spite of the dust-to-dust decree, early Christians found cremation distasteful. A few centuries after the birth of Christ burial became the fashionable method of body disposal in much of Europe.

Some of the Vikings, not accustomed to taking orders from Christians, continued to burn the occasional hero inside his boat. The Scandinavian countries are big on burning even today. And in non-Christian parts of the world, especially in Japan and India, cremation is the hands-down favorite.

It wasn't until the late 1800s that a handful of cremation devotees revived the practice in England and the United States. Slowly, retorts, or crematories, began to appear in both nations. And these days cremation in the United States is . . . well, it's hot.

About 25 percent of all the dead bodies in the United States are cremated. That's about half a million people. By 2010 the rate is expected to reach nearly 40 percent. The trend is unevenly spread. Those in the western third of the nation, along with New Englanders, are quick to render themselves dust. The middle of the country much prefers the slow underground route. Only 7 percent of Mississippians opt for cremation.

Bodies usually arrive at the crematorium in a rigid cardboard box. They are rarely embalmed. Some bodies, though, arrive from a traditional funeral service, for which they've been fully embalmed, made up, dressed up, and installed in a glossy metal or wood casket. However a body arrives, the retort operator usually slides the entire box into a furnace heated to between 1,400 and 1,800 degrees Fahrenheit. The cardboard box or wooden casket explodes in flame and disappears. The body goes more slowly.

Paul Lemieux is a chemical engineer and combustion expert at the EPA. He compares the next steps to what happens in backyards every day. "Most of what goes on is akin to cooking," he says. "The body's going to stay at about one hundred Celsius until all the water is driven off. When you cook hamburgers on a grill, the water's driven off; then the remaining material can heat up. The grease starts to burn. The body burns like regular organic fuel. It's not until you get very high temperatures that the bones and teeth burn."

Over the course of about an hour the bulk of the body is reduced to gases that are reburned in a special chamber and then sent up the stack. Specks of carbon soot might blacken the water vapor and other gases that boil from the body. Nitrogen oxide gas might tint the vapor plume faintly orange. The mercury in dental fillings vaporizes and swarms into the air. Burning fats evolve into complex hydrocarbons. As fire attacks the body's copious salt content, chlorine rises up the stack. As it cools, it has an opportunity to form tiny traces of dioxin.

The happy news, for those who wish to hasten their return to dust without fouling the atmosphere, is that cremation produces so little pollution that the EPA has deemed it a low priority for regulation. An air-pollution test at a Bronx, New York, crematory in 1999 once again begged a comparison to the homemaking arts—not to the barbecue grill, but to the fireplace.

A fireplace, the investigators reported, that sheds its orange cheer for an hour, can load the neighborhood air with nearly half a pound of particulates—that's airborne dusts of all flavors and sizes. A body that burns in an hour, however, produces just a bit more than half an *ounce* of particulates. Under the dirtiest conditions, with the crematory running at its hottest setting, each body produces about a quarter pound of microscopic dusts. The conclusion: when it comes to dusting the great outdoors, burning a body doesn't hold a candle to burning logs in the fireplace.

However, the bad news for those who yearn to burn is that the nation's 1,400 crematoria may each send about two pounds of toxic mercury into the air every year, from vaporized dental fillings. The average body contributes about a quarter gram of this malicious metal to the total. Those who wish to depart without poisoning the air might dictate that their fillings be removed before cremation. (Funeral directors routinely remove pacemakers before they send a body on to the crematory.)

As a cremation progresses, the body shrinks to a pile of bones, which shatter in the heat. When the crematory is turned off, 90-something percent of a person has departed to mix with the thick swirl of living and dead dusts that inhabit all outdoor air. Some of the "person particles" may settle slowly to Earth. Some of the gases may condense into beads that gather moisture from the atmosphere, forming raindrops. When these raindrops fall, they may sweep additional human dusts out of the air. So, many of a body's borrowed elements eventually return to Earth.

The unburnable elements lie on the floor of the furnace.

. . .

The "cremains" amount to six to ten pounds of bone, mixed with traces of hardy metals. The bone fragments are usually white or gray. As the operator scrapes them from the chamber, fine bits of the firebrick join the bone fragments. Crematory staff will run a magnet through the ashes to snare any metal bridgework, clothing fasteners, or surgically installed pins, plates, or joints.

A few crematories will turn over the remaining white grains and the long pieces of bone to the family. But most now use special grinders to pulverize all the cremains until the largest pieces are the size of coarse sand, or even finer. Why? The better to scatter them, of course.

In the past century, says Kenneth Iserson, "there have been very few *new* things that have been done with bodies—and even fewer new things done with ashes." But that doesn't mean people aren't trying. What started as a solemn ritual of scattering ashes off a mountain or into the ocean has evolved to include blasting bone dust into space and stuffing it into jewelry, fishing rods, greeting cards, and ceramic knickknacks. Throwing bone dust on sacred Aztec ruins threatened to become trendy in the Southwest, until the National Park Service banned scattering in many parks.

And there are more ashes to toss every year. A survey done by the Cremation Association of North America (CANA) found that in 1998 only about four in ten little boxes of bone dust were delivered to a cemetery. Those packets of dust were either buried, stored in an aboveground "columbarium," or emptied in a special "scattering garden."

But what became of the other hundreds of thousands of boxes of dust? Well, crematory staff themselves scattered about 64,000 boxes over water and another 24,000 boxes over land, in accordance with the wishes of the families. Almost 6 percent of all boxes of ashes were never picked up from crematories by family members, and remain in a dusty state of limbo.

The remaining 176,000 boxes of ashes? According to the CANA study these boxes or urns of bone dust were "taken home." But as to where the ashes went from there, the wildest guess probably wouldn't be too far out of line. And for those families who can't generate their own bright ideas, a plethora of companies now offer a panoply of ways to put dust to rest.

Creative Cremains, for instance, stirs flower seeds and a dash of ash into paper pulp, to make handmade paper cards. The recipients of these twenty-five-dollar missives are expected to cut them into pieces and plant them—bone dust, seeds, and all. The same San Francisco outfit offers to convert treasured figurines and other items into keepsake urns. (Whereas a special "scattering urn" is intended to beautify the act of scattering, a "keepsake urn" holds just a pinch of the departed and resides on the mantelpiece or bureau. "Keepsake pendants" are dust-stuffed jewelry items.)

A Claremont, California, company immortalizes the deceased by injecting a starburst of white bone ash into a heavy glass orb. Another company captures ashen dolphins inside a chunk of clear acrylic. For the sports-minded, a fellow in the Midwest has offered the service of packing cremains into shotgun shells and firing them at the game animal of choice—although he has happily packed the gun-shy into bowling balls, duck decoys, and even fish bait as well. And speaking of fish, a Georgia company will mix a batch of ashes into a batch of concrete. From this mixture it will cast mushroom-shaped artificial reef forms. These are sunk into the sea (bronze plaque optional) to attract coral and to shelter fish.

If your dear departed wished to be *scattered* and chose an inconvenient lo-cale, a burgeoning number of professional scatterers will now do that for you also. Various services will pour ashes off a sailboat or motorboat, out of an air-plane, into an Idaho forest, or in the Holy Land.

Throwing ashes out of an airplane may seem a bit staid in this ash-happy era. But it is a fine way to send fine dust on a long journey. The bulk of the bone bits will settle to the ground quickly. (The need to grind up bone fragments be-comes obvious.) But the finer dust could blow as far as wind and weather will allow. The very finest grains might soar for days, crossing oceans, traversing distant deserts, and effortlessly scaling exotic mountain ranges.

Fireworks give ashes a more dramatic entrance to the wind. For a few thou-sand dollars a Southern California company, Celebrate Life, will pack cre-mains into modified fireworks shells. Then, with friends and family gathered, the staff will set off both traditional and ash-packed fireworks, to music se-lected by the survivors. Like ashes scattered from an airplane, these will also enjoy the prospect of catching a sprightly wind.

Still too tame? Consider hurling those ashes into space. In 1997 the Celestis company launched its first load of cremains into orbit around the Earth. The Celestis ash capsule was piggybacked onto a disposable motor, whose main job was to lift a commercial rocket into orbit. When this motor burned out, it was cast aside, still carrying the Celestis capsule. And there, circling the planet in a relatively low orbit, remain the cremains of twenty-four pioneering clients. Or *part* of their cremains, anyway.

"About seven grams," says spokesman Christopher Pancheri. "We're a *memorialization* service." The company suggests that families do something else with the bulk of a space traveler's ashes.

That first batch of ashes, riding their rocket motor, will orbit the Earth until sometime in 2007. Then the whole kit and caboodle will sink too close to the Earth's sticky atmosphere and burn up—"like a shooting star," says Pancheri. The bone dust, along with the rocket motor, will be vaporized to swirl through the high atmosphere.

About 700 grams' worth of people have now taken the rocket ride—including that intrepid pioneer of *inner* space, Timothy Leary. Depending on the altitude of their particular rocket launch, some of these pinches of dust might circle the Earth for two centuries, making about fifteen loops a day. The cost: a bit under $1,000 per gram of dust.

But for Celestis, Earth orbit is a mere test drive. In 1998 the company cooperated with NASA to send a few dusty grams of the renowned comet scholar Eugene Schoemaker to the Moon. Forthcoming Moon shipment for the ashes of the masses will cost about $12,000 per customer. And Celestis is now taking reservations for a trip to the final frontier. In late 2001 Celestis will pack a pod of ashes on the *Encounter 2001* spacecraft. This non-NASA vehicle will saunter clear out of the solar system and into the void, carrying a payload of human hair, poetry, artwork, and bone dust.

Now, if the image of your cremains wandering through space, or even blowing around the planet in the company of desert dust, fungi, and soot, is off-putting, then mummification may be for you. Even in today's imaginative marketplace few services hold such promise in *preventing* your demotion to dust.

For about what it costs to have a few grams of your ashes exiled to the Moon, Summum, based in California—where else?—will do the job. The patented process preserves DNA—which is crucial if your long-term plans include cloning. "*I'd* only like to be cloned for the benefit of science," demurs Corky Ra, who developed the modern mummification method and who presides over the nonprofit company. "If they want to clone me when they've reached that point, they can."

To fend off your dusty fate for a *really* long time, splurge on Summum's $36,000 "mummiform." Both the sleek stainless-steel pod and the more traditional Egyptian-motif bronze shell are a quarter-inch thick. Before the wind and weather can begin abrading you into particles, they'll have to beat through the metal, erode the egg of synthetic amber that fills the mummiform, and shred your wrappings. So far Ra has successfully—and rather elegantly—mummified pets as well as thirty human cadavers at a medical school. And while Summum has yet to preserve a paying person, more than a hundred people have paid in advance for the service, which can be performed at many funeral homes.

And how long will mummification fend off your dusty fate? Corky Ra sighs at the question. The Bronze Age was too recent to provide good data. "Four-thousand-, five-thousand-year-old bronzes are known," he says.

But even if your mummiform lasts 100,000 years, on the Earth's clock, that's the blink of an eye. You *will* be dust, in time. If by some freak chance you should find a way to elude wind and water, not to mention the grinding of the tectonic plates that steadily recycle the Earth's very crust, it still must be:

You will be dust.

Indeed, the entire *Earth* will be dust. When the Sun's central furnace has converted its supply of hydrogen atoms to helium atoms, the star will heat up, swell up, and become a "red giant." Today, if the Sun were a small grape, the Earth would be a grain of sand orbiting five feet away. But as the Sun goes red giant, that grape will balloon right out to graze our sand speck.

And although earlier calculations suggested that the Earth would back away from this menace, Lee Anne Willson no longer thinks that will be the case. Willson, an Iowa State University physicist and astronomer, made a media splash in early 2000 when she presented her peers with the gloomy predictions of her computer model.

"I've never done anything as popular as roasting the Earth," admits the chipper Willson.

Some previous models had predicted that the geriatric Sun will shed huge amounts of gas into space when it starts to balloon outward. The Sun's loss of mass would weaken its gravitational hold on the Earth, which would slide out to a safer orbit.

"Mass loss makes all the difference in the world, in whether the Earth becomes dust or remains a little nugget memorial to mankind," says Willson. But her calculations show that even a scorched nugget is more than we can hope for.

Not that we'll be around to wring our prehensile hands over this slight. Even before the Sun's thin atmosphere expands to the Earth's orbit, conditions on the planet will get mighty unhealthful, and then extremely dusty. That's not due to our own polluting habits, but because the Sun naturally grows a little bit hotter every day.

Ken Caldeira is a research scientist at Lawrence Livermore National Laboratory. He also uses a computer model to read the planet's fortune. But he concentrates on the eons between now and when Lee Anne Willson roasts it. In just a billion years or so from today, Caldeira predicts, the steady heating of the planet will have totally transformed the chemistry of the atmosphere. All the plants will have withered and died, leading the slow parade of living things to the grave. We won't be around for the really dusty part.

"In about a billion and a half years, as the Earth heats up, more water vapor will enter the atmosphere from the oceans," Caldeira predicts. This evaporated water won't be satisfied to stay down in the wet and weathery troposphere. It will migrate upward, he says, to join the thin gases of the stratosphere above.

"When the water vapor gets up into the stratosphere, it gets bombarded by solar radiation, which will break up the molecules," Caldeira continues. "And the hydrogen atoms will get so energized that they can leave the planet—forever. And once you lose the water? My guess is the Earth becomes a rather dusty place."

Perhaps some hardy bacteria will be able to keep a foothold on the barren ball of warm dust and rock that will remain. But even they won't escape a dusty end. A couple of billion years after the oceans dry up, the Sun's expanding atmosphere will approach the Earth like a boiling red wall, burning at perhaps 6,000 degrees Fahrenheit. And just as an orbiting rocket motor is gradually slowed by the Earth's outermost atmosphere, the Earth's pace around this looming Sun will slow as it fights the halo of gas surrounding the red giant.

"The Earth will spiral in," says Lee Anne Willson. "It will get hotter and hotter as it approaches the Sun's opaque interior. And then once it's in, it will vaporize."

That vapor of the Earth's quartz and granite, iron and gold, its fresh and fossilized bones, and any stainless-steel mummiforms that might be lying about, will swirl in the red giant's atmosphere. The deserts and mountains, the blood-darkened soils and ancient oil deposits—all will burn to gases. Every dust bunny and every vacuum cleaner will flash away in the red-hot crematory of the Sun.

And then those strange vapors may form fresh dust.

"Near the end of a red giant's life it's a big fluffy thing that expands and contracts like a beating heart, in a cycle that usually takes a little less than a year," says Willson. The slow throbbing produces enormous shock waves that travel out through the star's atmosphere.

"When a shock wave compresses the gas, it heats up. But then when the gas expands again, it cools, and dust can form. That dust survives the next heating phase, and during each cooling phase it grows."

Some of the vapors of our planet will be allowed to recondense in the atmosphere of this beating heart. Minuscule grains of fluffy dusts, chemically akin to quartz rocks and nickel-iron, will grow.

Then, driven on the solar wind, this puff of Earth smoke will blow out into the galaxy. Shortly thereafter even the Sun will shudder its outer layers prettily into space, where a fraction of the glowing gases will also cool into simple dusts.

Now, it must be noted that there's a chance the Earth will, in fact, escape both the Sun's grip and its death throes. Fred Adams, a University of Michigan physicist and a former student of Willson's, has described some of our planet's options in a delightful book called *The Five Ages of the Universe*. One alternative, he says, is that a close-passing red dwarf star could gravity-whip the Earth out of orbit and into deep, cold space. The good news is that this would prolong the planet's time in the universe by about 10 trillion trillion trillion years, Adams says. The bad news is that these years would be frigid and lonesome ones, and in the end the planet would evaporate into space, through a subtle, subatomic process called "proton decay." Adams has calculated the odds of this happening within the next 2 billion years.

"About one in one hundred thousand. You wouldn't bet real money on those odds," he adds, laughing. "But it's actually better than most lottery tickets."

Adams's rosier vision has bleaker odds. If a pinwheeling *pair* of red dwarf stars were to wobble through the neighborhood, the Earth would stand a chance of being forcibly adopted into this family. Red dwarf stars burn in a cool and miserly manner. So this adoption might keep the Earth safe and warm for trillions of years—thousands of times longer than if it sticks with our own hot Sun. In this scenario the Earth again ends with a whimper of sub-atomic evaporation. The odds of this alternative? One in 3 million.

"The *best* way to make dust," Adams says, "would be to have the Earth eaten by the Sun." And the odds of this scenario, Willson and Adams agree, are excellent.

A human body that's fed to vultures continues to wander through the broad web of life, being eaten, then expelled, eaten and expelled. And the remains of our Earth will probably do likewise.

We've been granted a 10-billion-year lease on the space dust that makes up our solar system. But even when that lease expires, the universe will still be in its tender infancy. The dust we borrowed from it will enjoy many more incar-nations.

Billions of years after our dust drifts out into the daunting distances of the galaxy, it might find itself stirred into a dark cloud that's gathering around a star seed. A smattering of our dust may find itself whirled into the new star's very core. A bit more might be rolled into a planet that circles that star. And then, if the star is a big one, it might promptly explode, spattering both old and new-made dust back into the galaxy.

And so, with each generation of stars, the universe will grow dustier. As the trillions of years flow by, the night sky will darken as dust blots out more starlight. The stars themselves will burn more coolly and shine more dimly as they adapt to this dustier fuel. As the universe ages, Fred Adams foresees a generation of bizarre stars so enriched with insulating dust that their atmospheres will swirl with ice crystals.

And then, like an old newspaper in the attic, the worn-out universe will gradually disappear under the thickening dust.

WEB SITES

2: Life and Death among the Stars

A gorgeous portrait of our galaxy, the Milky Way, including lots and lots of dust: http://www.star.ucl.ac.uk/~apod/apod/ap980128.html

An illustrated diary of the Earth's formation in the original dusty disk: http://www.psi.edu/projects/planets/planets.html

Infrared technology lets astronomers see through dust as though they had X-ray vision. This site lets you see IR in action: http://www.ipac.caltech.edu/Outreach/Edu

A stereogram of interstellar dust: http://www.astro.ucla.edu/~wright/dust/

Stunning photos of the zodiacal light, with comet Hale-Bopp thrown in for extra drama: http://educeth.ethz.ch/stromboli/photos/photocom/index-e.html

Did magical molecules on space dust deliver life to Earth? The Astrochemistry Lab at NASA's Ames Research Center has research links and recent articles on the subject: http://web99.arc.nasa.gov/~astrochm/

Astrobiology, or the study of life among the stars (including ours), is a booming new field of research. NASA's site includes news, interviews, Q&A, and feature stories: http://.astrobiology.arc.nasa.gov

This reader-friendly article about how life's raw materials are made in space includes good illustrations and was written by scientist Max Bernstein and colleagues: http://www.sciam.com/1999/0799issue/0799bernstein.html

3: A Light and Intriguing Rain of Space Dust

Stardust, the mission to trap comet dust and return it to Earth, has its own Web site: http://stardust.jpl.nasa.gov/mission/msnover.html

NASA's dust-collecting division has a rich site with gorgeous portraits of the little specks: http://www-curator.jsc.nasa.gov

Comets come to life at the Web site of astronomer David Jewitt, a comet expert from the University of Hawaii: http://www.ifa.hawaii.edu/faculty/jewitt/kb.html

Asteroids have their day at this Lunar and Planetary Laboratory site: http://seds.lpl.arizona.edu/nineplanets/nineplanets/asteroids.html

Did cosmic dust cause the ice ages? Berkeley professor Richard A. Muller's site presents his research on this subject—and on his theory that our Sun has an unseen companion star: http://muller.lbl.gov/

Larry Nittler, who studies ancient dust by digging it out of meteorites, maintains a nice Web site—including an image of space diamonds! http://www.ciw.edu/lrn/psg_main.html

4: THE (DEADLY) DUST OF DESERTS

Hear Woody Guthrie sing "The Dust Pneumonia Blues" at: http://chnm.gmu.edu/courses/hist409/dust/dust.html

Wind erosion didn't blow away with the Dust Bowl years. The problem continues, as the Wind Erosion Research Unit can testify: http://www.weru.ksu.edu/

Dust on the wind, and its scouring effect on deserts and rocks, is explored with wonderful photographs at this USGS site: http://pubs.usgs.gov/gip/deserts/eolian/

Portraits of *Oviraptors*—and many other dinosaurs—can be found at: http://web.syr.edu/~dbgoldma/pictures.html

"Earth from Space" is NASA's gallery of smashing shuttle photographs, including dust storms in the Taklamakan Desert of China and the Djourab Sand Region of Chad: http://earth.jsc.nasa.gov/

5: A STEADY UPWARD RAIN OF DUST

The Pan American Aerobiology Association publishes its newsletter and conference summaries on the Internet. For the very latest on rambling molds and marauding pollens, go to: http://www.paaa.org

NASA's home-looking division publishes a wealth of information on fires and other global issues—with photos: http://earthobservatory.nasa.gov/

The Meteor Crater catastrophe in Arizona didn't kill any dinosaurs, but it's a good example of the damage an incoming asteroid can do: http://www.barringercrater.com

This site is dedicated to ancient pollen, spores, and other microscopic fossils. "Pollen grain of the month" and children's corner included: http://www.geo.arizona.edu/palynology

Volcanoes are the stars at this United States Geological Survey site, which includes links to various volcano observatories: http://vulcan.wr.usgs.gov/home.html

Smoke from wildfires sometimes forms such vast plumes that it's easily visible from space. A series of astronaut photos is at: http://eol.jsc.nasa.gov/newsletter/smoke/page1.html

Diatoms, in stunning detail, are cataloged at this site: http://www.bgsu.edu/departments/biology/algae/html/Image_Archive.html

6: Dust on the Wind Heeds No Borders

A massive dust storm swirls out of Asia, crosses the Pacific, and blurs the western United States in this series of satellite photographs: http://daac.gsfc.nasa.gov/CAMPAIGN_DOCS/OCDST/asian_dust.html

The dust storm now called "The Asian Dust Event of April 1998" is re-created, animated, discussed, and explained by interested scientists at this rich site: http://capita.wustl.edu/Asia-FarEast/

The National Oceanic and Atmospheric Administration maintains a site that features satellite imagery of what they call "significant events." These range from eclipses to dust storms and giant plumes of smoke: http://www.osei.noaa.gov

Weather Modification, Inc.'s, Web site has lots of information on how cloud seeding works, including diagrams of hail-making clouds: http://www.wmi.cban.com/services.html

Deserts and other landforms as photographed from space, with an emphasis on the changing environment are at: http://edcwww.cr.usgs.gov/earthshots/slow/tableofcontents

7: Did Dust Do In the Ice Age?

The National Ice Core Lab collects and distributes ice cores to scientists. The "how it is done" section of this site has a chilling slide show of the entire ice-harvesting process. Dress warmly: http://nicl.usgs.gov/index.html

"Rapid Climate Change," an *American Scientist* article by Kendrik Taylor, details recent discoveries about the dizzying speed with which the world's temperature can rise—or fall: http://www.maxey.dri.edu/WRC/waiscores/Amsci/Taylor.html

This summarizes work by the USGS to understand how climate and dust interact in the southwestern United States. Includes a stunning aerial photo of a dust storm in the San Joaquin Valley. And this page is just one in a special series of meaty features on USGS projects and discoveries: http://geochange.er.usgs.gov/sw/impacts/geology/dust/

The EPA's pages on global climate change are simple and straightforward and include a "what can I do?" component: http://www.epa.gov/globalwarming/

Daniel Rosenfeld's satellite-aided method of locating "pollution tracks" is described here. Includes links and color-coded images of the tracks: http://earthobservatory.nasa.gov/Study/Pollution/

8: A Steady *Downward* Rain of Dust

Some of dust's most ardent admirers advocate sprinkling the stuff in the garden for bigger, better produce. Great history, testimonials: http://Remineralize-the-Earth.org

Hot shots on this NASA site include dust storms, giant smoke plumes, and other large-scale phenomena. Follow links from: http://www.gsfc.nasa.gov

The story of Saharan dust and Caribbean coral reefs is told with text and many photos at: http://coastal.er.usgs.gov/african_dust/

Lungs, lungs, lungs, and lung diseases are the subject of the American Lung Association Web site: http://www.lungusa.org/

The United Nations Environmental Program takes a global view of long-lived and high-flying pollutants, POPs included. Zillions of links and tons of information: http://irptc.unep.ch/

A useful article about the dangers of supersmall dusts, with a nice diagram of lung tissue, can be downloaded from this site. The article is called "Small Particles—Big Problem." http://www.tsi.com/hsi/homepage/applnote/iti_067.pdf

9: A Few Unsavory Characters from the Neighborhood

This site has pretty pictures of Cappadocia's fabulous caves, including frescoed churches, houses, and the astonishing "underground cities" that ramble for black and mysterious miles: http://www.hitit.co.uk/regions/cappy/About.html

Learn what industries add to your hometown dust through the EPA's Toxic Release Inventory program. The "Explorer" tool allows you to dig through data county by county: http://www.epa.gov/tri

Is your workplace dusty? The National Institute of Occupational Safety and Health site has loads of dust-at-work research: http://www.cdc.gov/niosh/homepage.html

NIOSH's "Work-Related Lung Disease Surveillance Report, 1999," tracks the grim numbers of death and disease in various dusty professions: http://www.cdc.gov/niosh/w99cont.html

How many people die of . . . whatever! Check out the Centers for Disease Control site for endless descriptions of diseases, death rates, and other statistics. Don't miss the "hoaxes and rumors" page: http://www.cdc.gov

Arguably the world's greatest pathology site, containing everything you ever wanted to know about disease and death, including detailed descriptions and photographs. The pages are colossal, slow to load, and worth waiting for: www.pathguy.com/index1.htm

10: Microscopic Monsters and Other Indoor Devils

The Environmental Protection Agency and the Consumer Product Safety Commission's publication on indoor air pollutants, *The Inside Story: A Guide to Indoor Air Quality,* is at: http://www.cpsc.gov/cpscpub/pubs/450.html

How many bug legs are allowed in your flour? The Food and Drug Administration's purity standards for everything from apricots to cornmeal: http://vm.cfsan.fda.gov/~dms/dalbook.html

A stunning "mite site," with fabulous color photos of some of the world's most beautiful and bizarre mites: http://www.uq.edu.au/entomology/mite/mitetxt.html

The American Lung Association of Washington State has a good site for indoor-air-pollution issues: http://www.alaw.org

Asthma is explained at the National Heart, Lung and Blood Institute site: http://www.nhlbi.nih.gov/health/public/lung/index.htm

How is an asthmatic lung different from a normal one? Check this diagram: http://www.asthmacentre.com/manual/howasthma.html

The American Academy of Asthma, Allergy and Immunology also dives deep into asthma: http://www.aaaai.org/public/default.stm

11: Dust to Dust

Statistics on which states in the United States do the most cremation, and what happens to the ashes are among the publications at the Cremation Association of North America site: http://www.cremationassociation.org

The Internet Cremation Society has links to information, plus urns, scattering services, and other the postlife essentials: http://www.cremation.org

Whether you need an acrylic dolphin ash holder, an urn named "Majestic," or ideas about where to scatter those ashes, this is the all-but-one-stop shop for cremation fans: http://www.urnmall.com

As for that other stop, drop by the Neptune Society's Web site and reserve your spot in a crematory near you. You can e-purchase your own roasting—and a scattering at sea—ahead of time: http://www.neptunesociety.com

The cremains-to-coral company is at: http://www.eternalreefs.com

The cremains-in-space company is at: http://www.celestis.com

For the burial-minded, a site where you might order your casket, vault, and marker: http://www.thefuneralstore.net

For the mummification-minded, the Summum site is a rambling collection of New Age spirituality, music, plus mummification photos and prices: http://www.summum.org

When our Sun flings off its outerwear, the explosion won't be big enough to win supernova status. Rather, a pretty flower of glowing gas called (inappropriately) a "planetary nebula" will bloom in the sky. The Hubble Space Telescope has captured some of these, which are displayed at: http://oposite.stsci.edu/pubinfo/pr/97/38/b-js.html

BIBLIOGRAPHY

1: THE WORLD IN A GRAIN OF DUST

Cooke, William F., et al. "A Global Black Carbon Aerosol Model." *Journal of Geophysical Research,* 101, no. D14, pp. 19,395–19,419, 1996.

EDGAR Database. "Global Anthropogenic NOx Emissions in 1990." Published at: rivm.nl/env/int/coredata/edgar/

Ford, A., et al. "Volcanic Ash in Ancient Maya Ceramics of the Limestone Lowlands: Implications for Prehistoric Volcanic Activity in the Guatemala Highlands." *Journal of Volcanology and Geothermal Research,* 66, no. 1-4, pp. 149-162, 1995.

Gong, Sunling. Global Sea-Salt Flux Estimate. Personal communication, January 2000.

Guenther, Alex. Biogenic Volatile Organic Compounds, Global Flux Estimates. Personal communication, January 2000.

Kaiser, Jocelyn. "Panel Backs EPA and 'Six Cities' Study." *Science,* 289, p. 711, August 4, 2000.

Marshall, W. A. "Biological Particles over Antarctica." *Nature,* 383, p. 680, October 24, 1996.

Prospero, Joseph M. "Long-Term Measurements of the Transport of African Mineral Dust to the Southeastern United States: Implications for Regional Air Quality." *Journal of Geophysical Research,* 104, no. D13, pp. 15,917–15,927, 1999.

Psenner, R., et al. "Life at the Freezing Point." *Science,* 280, pp. 2,073–2,074, June 26, 1998.

Sattler, B., et al. "Bacterial Growth in Supercooled Cloud Droplets." *Geophysical Research Letters,* 28, no. 2, pp. 239-243, 2001.

Stone, E. C., et al. "From Shifting Silt to Solid Stone: The Manufacture of Synthetic Basalt in Ancient Mesopotamia." *Science,* 280, pp. 2,091–2,093, June 26, 1998.

Tegen, Ina, et al. "Contribution of Different Aerosol Species to the Global Aerosol Extinction Optical Thickness: Estimates from Model Results." *Journal of Geophysical Research,* 102, no. D20, pp. 23,895–23,915, 1997.

Urquart, Gerald, et al. "Tropical Deforestation." NASA Earth Observatory, un-

dated. Published at: earthobservatory.nasa.gov/Library/Deforestation/deforestation
_3.html

U.S. Centers for Disease Control. *Work-Related Lung Disease Surveillance Report 1999.*
Washington, D.C.: CDC, 1999.

Yokelson, Robert J. Gas-to-Particle Conversion Rate for Biomass-Burning Carbon.
Personal communication, January 2000.

2: LIFE AND DEATH AMONG THE STARS

Andersen, Anja, et al. "Spectral Features of Presolar Diamonds in the Laboratory and in Carbon Star Atmospheres." *Astronomy and Astrophysics,* 30, pp. 1,080–1,090,
1998.

Backman, Dana, et al. "Extrasolar Zodiacal Emission: NASA Panel Report." NASA,
1997. Published at: http://astrobiology.arc.nasa.gov/workshops/1997/zodiac/back-
man/backman/IIIa2.html

Basiuk, Vladimir A., et al. "Pyrolytic Behavior of Amino Acids and Nucleic Acid
Bases: Implications for Their Survival During Extraterrestrial Delivery." *Icarus,* 134, no.
2, pp. 269–279, 1998.

Beckwith, Steven V. W., et al. "Dust Properties and Assembly of Large Particles in
Protoplanetary Disks." From *Protostars and Planets IV.* Mannings, Vince, et al., eds. Tucson: University of Arizona Press, 2000.

Bernstein, Max P., et al. "Life's Far-Flung Raw Materials." *Scientific American,* 281,
pp. 42–49, July 1999.

Blum, Jurgen, et al. "The Cosmic Dust Aggregation Experiment CODAG." *Measurement Science and Technology,* 10, pp. 836–844, 1999.

Clark, David H. *The Historical Supernovae.* Oxford: Pergamon Press, 1979.

Clayton, Donald D., et al. "Condensation of Carbon in Radioactive Supernova
Gas." *Science,* 283, pp. 1,290–1,292, February 26, 1999.

Culotta, Elizabeth, et al. "Planetary Systems Proliferate." *Science,* 286, p. 65, October
1, 1999.

Dwek, E., et al. "Detection and Characterization of Cold Interstellar Dust and Poly-
cyclic Aromatic Hydrocarbon Emission, from COBE Observations." *Astrophysical Journal,* 475, pp. 565–579, February 1, 1997.

Hellmans, Alexander. "Fine Details Point to Space Hydrocarbons." *Science,* 287,
p. 946, February 11, 2000.

Irion, Robert. "Can Amino Acids Beat the Heat?" *Science,* 288, p. 605, April 20, 2000.

Lada, Charles. "Deciphering the Mysteries of Stellar Origins." *Sky & Telescope,*
pp. 18–24, May 1993.

Maran, Stephen P., ed. *The Astronomy and Astrophysics Encyclopedia.* New York: Van
Nostrand Reinhold, 1992.

Mathis, John S. "Interstellar Dust and Extinction." *Annual Review of Astronomy and
Astrophysics,* 28, no. 28, pp. 37–69, 1990.

Reach, William T., et al. "The Three-Dimensional Structure of the Zodiacal Dust
Bands." *Icarus,* 127, no. 2, pp. 461–485, 1997.

Stokstad, Erik. "Space Rock Hints at Early Asteroid Furnace." *Science,* 284, pp. 1,246–1,247, May 21, 1999.

Wood, John A. "Forging the Planets." *Sky & Telescope,* pp. 36–48, January 1999.

3: A LIGHT AND INTRIGUING RAIN OF SPACE DUST

Andersen, Anja, et al. "Spectral Features of Presolar Diamonds in the Laboratory and in Carbon Star Atmospheres." *Astronomy and Astrophysics,* 30, pp. 1,080–1,090, 1998.

Backman, Dana, et al. "Extrasolar Zodiacal Emission: NASA Panel Report." NASA, 1997. Published at: astrobiology.arc.nasa.gov/workshops/1997/zodiac/backman/backman/IIIa2.html

Bradley, John P., et al. "An Infrared Spectral Match between GEMS and Interstellar Grains." *Science,* 285, pp. 1,716–1,718, September 10, 1999.

Farley, K. A. "Cenozoic Variations in the Flux of Interplanetary Dust Recorded by 3He in a Deep-Sea Sediment." *Nature,* 376, pp. 153–156, July 13, 1995.

———, et al. "Geochemical Evidence for a Comet Shower in the Late Eocene." *Science,* 280, pp. 1,250–1,253, May 22, 1998.

Haggerty, Stephen E. "A Diamond Trilogy: Superplumes, Supercontinents, and Supernovae." *Science,* 285, pp. 851–860, August 6, 1999.

Kerr, Richard A. "Planetary Scientists Sample Ice, Fire, and Dust in Houston." *Science,* 280, pp. 38–39, April 3, 1999.

Kortenkamp, Stephen J. "Amid the Swirl of Interplanetary Dust." *Mercury,* pp. 7–11, November–December 1998.

———, et al. "A 100,000-Year Periodicity in the Accretion Rate of Interplanetary Dust." *Science,* 280, pp. 874–876, May 8, 1998.

Kyte, Frank T. "The Extraterrestrial Component in Marine Sediments: Description and Interpretation." *Paleoceanography,* 3, no. 2, pp. 235–247, 1988.

Love, S. G., et al. "A Direct Measurement of the Terrestrial Mass Accretion Rate of Cosmic Dust." *Science,* 262, pp. 550–553, October 22, 1993.

Maurette, M., et al. "A Collection of Diverse Micrometeorites Recovered from 100 Tonnes of Antarctic Blue Ice." *Nature,* 351, pp. 44–45, May 2, 1991.

———. "Placers of Cosmic Dust in the Blue Ice Lakes of Greenland." *Science,* 233, pp. 869–872, August 22, 1986.

Monastersky, Richard. "Space Dust May Rain Destruction on Earth." *Science News,* 153, no. 19, p. 294, May 9, 1998.

Muller, Richard A., et al. "Origin of the Glacial Cycles: A Collection of Articles." International Institute for Applied Systems Analysis, RR-98-2, February 1998.

Murray, John, et al. "Report on Deep-Sea Deposits Based on the Specimens Collected During the Voyage of *HMS* Challenger in the Years 1872–1876. Volume XVIII (Part 1), pp. xcix–c. From *The Voyage of HMS* Challenger. Thompson, C. W., and Murray, J., eds. London: Her Majesty's Stationery Office, 1891.

Oliver, John P., et al. "LDEF Interplanetary Dust Experiment (IDE) Impact Detector

Results." Paper presented at the SPIE International Symposium on Optical Engineering in Aerospace Sensing, April, 1994.

Taylor, Susan, et al. "Accretion Rate of Cosmic Spherules Measured at the South Pole." *Nature*, 392, pp. 899–903, April 30, 1998.

4: The (Deadly) Dust of Deserts

Anonymous. "Some Information about Dust Storms and Wind Erosion on the Great Plains." U.S. Department of Agriculture, Soil Conservation Service. March 30, 1953. (AGR-SCS-Beltsville, Maryland 2630, April 1954)

Babaev, Agajan G., ed. *Desert Problems and Desertification in Central Asia.* Heidelberg: Springer, 1999.

Bennett, H. H. "Emergency and Permanent Control of Wind Erosion in the Great Plains." *The Scientific Monthly*, XLVII, pp. 381–399, 1938.

———. *Soil Conservation.* New York and London: McGraw-Hill, 1939.

Blouet, Brian W. et al., eds. *The Great Plains: Environment and Culture.* Lincoln: University of Nebraska Press, 1979.

Busacca, Alan, et al. "Effect of Human Activity on Dustfall: A 1,300-Year Lake-Core Record of Dust Deposition on the Columbia Plateau, Pacific Northwest U.S.A." *Conference Proceedings: Dust Aerosols, Loess Soils & Global Change, Washington State University.* Publication No. MISC0190, 1998.

———, eds: *Conference Proceedings: Dust Aerosols, Loess Soils & Global Change, Washington State University.* Publication No. MISC0190, 1998.

Cloudsley-Thompson, J. L. *Man and the Biology of Arid Zones.* Baltimore: University Park Press, 1977.

———, ed. *Sahara Desert.* New York: Pergamon Press, 1984.

Crowley, Thomas J. "Remembrance of Things Past: Greenhouse Lessons from the Geologic Record." *Consequences*, 2, no. 1, pp. 3–12, 1996.

Douglas, David. "Environmental Eviction: Migration from Environmentally Damaged Areas." *Christian Century*, 113, no. 26, pp. 839–841, 1996.

Fastovsky, David E., et al. "The Paleoenvironments of Tugrikin-Shireh (Gobi Desert, Mongolia) and Aspects of the Taphonomy and Peleoecology of Protoceratops (Dinosauria: Ornithishichia)." *Palaios*, 12, no. 1, pp. 59–70, 1997.

George, Uwe. *In the Deserts of This Earth.* San Diego: Harcourt Brace Jovanovich, 1977.

Gillette, Dale A. "Estimation of Suspension of Alkaline Material by Dust Devils in the United States." *Atmospheric Environment*, 24A, no. 5, pp. 1,135–1,142, 1990.

Graham, Stephan A., et al. "Stratigraphic Occurrence, Paleoenvironment, and Description of the Oldest Known Dinosaur (Late Jurassic) from Mongolia." *Palaios*, 12, no. 3, pp. 292–297, 1997.

Helms, Douglas, et al., eds. *The History of Soil and Water Conservation.* Washington, D.C.: The Agricultural History Society, 1985.

Hendrix, Marc S., et al. "Noyon Uul Syncline, Southern Mongolia: Lower Mesozoic Sedimentary Record of the Tectonic Amalgamation of Central Asia." *GSA Bulletin*, 108, no. 10, pp. 1,256–1,274, 1996.

———. "Sedimentary Record and Climatic Implications of Recurrent Deformation in the Tian Shan: Evidence from Mesozoic Strata of the North Tarim, South Junggar, and Turpan Basins, Northwest China." *GSA Bulletin,* 104, pp. 53-79, January 1992.

Holden, Constance, editor. "Remnant Crocs Found in Sahara." *Science,* 287, p. 1,199, February 18, 2000.

Hurt, R. Douglas. *The Dust Bowl: An Agricultural and Social History.* Chicago: Nelson-Hall, 1981.

Jerzykiewicz, T., et al. "Djadokhta Formation Correlative Strata in Chinese Inner Mongolia: An Overview of the Stratigraphy, Sedimentary Geology, and Paleontology and Comparisons with the Type Locality in the Pre-Altai Gobi." *Canadian Journal of Earth Science,* 30, pp. 2,180-2,195, 1993.

Kerr, Richard A. "The Sahara Is Not Marching Southward." *Science,* 281, pp. 633-634, July 31, 1998.

Loope, David B., et al. "Life and Death in a Late Cretaceous Dune Field, Nemegt Basin, Mongolia." *Geology,* 26, no. 1, pp. 27-30, 1998.

———. "Mud-field *Ophiomorpha* from Upper Cretaceous Continental Redbeds of southern Mongolia: An Ichnologic Clue to the Origin of Detrital, Grain-Coating Clays." *Palaios,* 14, pp. 451-458, 1999.

Louw, G. N., et al. *Ecology of Desert Organisms.* New York: Longman Group, 1982.

Lumpkin, Thomas A., et al. "The Critical Role of Loess Soils in the Food Supply of Ancient and Modern Societies." *Conference Proceedings: Dust Aerosols, Loess, Soils & Global Change, Washington State University,* 1998.

Malusa, Jim. "Silent Wild. (Atacama Desert, Chile)." *Natural History,* 107, no. 3, pp. 50-57, 1998.

Priscu, John C. *Ecosystem Dynamics in a Polar Desert: The McMurdo Dry Valleys, Antarctica.* Washington, D.C.: American Geophysical Union, 1998.

Pye, Kenneth. *Aeolian Dust and Dust Deposits.* New York: Harcourt Brace Jovanovich, 1987.

Reheis, M. C., et al. "Owens (Dry) Lake, California: A Human-Induced Dust Problem." United States Geological Survey. 1997. Published at: http://geochange.er.usgs.gov/sw/impacts/geology/owens/

Sidey, Hugh. "Echoes of the Great Dust Bowl." *Time,* p. 50, June 10, 1996.

Sincell, Mark. "A Wobbly Start for the Sahara." *Science,* 285, p. 325, July 16, 1999.

Sletto, Bjorn. "Desert in Disguise." *Earth,* 6, no. 1, p. 42-50, 1997.

Sneath, David. "State Policy and Pasture Degradation in Inner Asia." *Science,* 281, pp. 1,147-1,148, August 21, 1998.

Strauss, Evelyn. "Wringing Nutrition from Rocks." *Science,* 288, p. 1,959, June 16, 2000.

U.S. Department of Agriculture. "Summary Report 1997 National Resources Inventory." Published at: www.nhq.nrcs.usda.gov/NRI

Walker, A. S. "Deserts: Geology and Resources." USGS, 1997. Published at: http://pubs.usgs.gov/gip/deserts/

5: A Steady Upward Rain of Dust

Anderson, Bruce E., et al. "Aerosols from Biomass Burning Over the Tropical South Atlantic Region: Distributions and Impacts." *Journal of Geophysical Research,* 101, no. D19, pp. 24,117–24,137, 1996.

Andres, R. J., et al. "A Time-Averaged Inventory of Subaerial Volcanic Sulfur Emissions." *Journal of Geophysical Research,* 103, no. D19, pp. 25,251–25,261, 1998.

Bates, Timothy, et al. "Oceanic Dimethylsulfide (DMS) and Climate." Date unknown. Published at: saga.pmel.noaa.gov/review/dms_climate.html

Baxter, P. J., et al. "Preventive Health Measures in Volcanic Eruptions." *American Journal of Public Health,* 76 (Suppl. 3), pp. 84–90, 1986.

Casadevall, Thomas J. "The 1989–1990 Eruption of Redoubt Volcano, Alaska: Impacts on Aircraft Operations." *Journal of Volcanology and Geothermal Research,* 62, pp. 301–316, 1994.

———, ed. *Volcanic Ash and Aviation Safety: Proceedings of the First International Symposium on Volcanic Ash and Aviation Safety.* USGS Bulletin 2047, 1994.

Chen, Jen-Ping. "Particle Nucleation by Recondensation in Combustion Exhausts." *Geophysical Research Letters,* 26, no. 15, pp. 2,403–2,406, 1999.

Dacey, John W. H., et al. "Oceanic Dimethylsulfide: Production During Zooplankton Grazing on Phytoplankton." *Science,* 233, pp. 1,314–1,316, September 19, 1986.

Ferek, Ronald J., et al. "Measurements of Ship-Induced Tracks in Clouds off the Washington Coast." *Journal of Geophysical Research,* 103, no. D18, pp. 23, 199–23, 206, 1998.

Friedl, Randall R., ed. *Atmospheric Effects of Subsonic Aircraft: Interim Assessment Report of the Advanced Subsonic Technology Program.* Goddard Space Flight Center: NASA Reference Publication 14-00, 1997.

Galanter M., et al. "Impacts of Biomass Burning on Tropospheric CO, NOx, and O3." *Journal of Geophysical Research,* 105, no. D5, pp. 6,633–6,653, 2000.

Herring, David. "Evolving in the Presence of Fire." Published at: earthobservatory.nasa.gov/Study/BOREASFire/boreas_fire.html. October, 1999.

Hornberger, B., et al. "Measurement of Tire Particles in Urban Air." Presented at ACAAI, Dallas, Texas, 1995.

Jaffrey, S. A., et al. "Fibrous Dust Release from Asbestos Substitutes in Friction Products." *Annals of Occupational Hygiene,* 36, no. 2, pp. 173–181, 1992.

Knight, Nancy C., et al. "Some Observations on Foreign Material in Hailstones." *Bulletin of the American Meteorological Society,* 59, no. 3, pp. 282–286, 1978.

Kuhlbusch, Thomas A. J. "Black Carbon and the Carbon Cycle." *Science,* 280, pp. 1,903–1,904, June 19, 1998.

Liss, Peter. "Take the Shuttle—from Marine Algae to Atmospheric Chemistry." *Science,* 285, pp. 1,217–1,218, August 20, 1999.

Marshall, W. A. "Biological Particles Over Antarctica." *Nature,* 383, p. 680, October 24, 1996.

McGee, Kenneth A., et al. "Impacts of Volcanic Gases on Climate, the Environment, and People." U.S. Geological Survey Open-File Report 97-262, 1997.

Newhall, Chris, et al. "The Cataclysmic 1991 Eruption of Mount Pinatubo, Philippines." U.S. Geological Survey Fact Sheet 113-97, online version 1.0. Published at: geo pubs.wr.usgs.gov/fact-sheet/fs113-97/

Nyberg, F., et al. "Urban Air Pollution and Lung Cancer in Stockholm." *Epidemiology*, 11, no. 5, pp. 487–495, 2000.

Penner, Joyce E., et al., eds. *Aviation and the Global Atmosphere*. Cambridge: Cambridge University Press, 1999.

Psenner, R. et al. "Life at the Freezing Point." *Science*, 280, pp. 2,073-2,074, June 26, 1998.

Pyne, Stephen J. *World Fire: The Culture of Fire on Earth*. New York: Henry Holt, 1997.

Quinn, P. K., et al. "Aerosol Optical Properties in the Marine Boundary Layer During the First Aerosol Characterization Experiment (ACE 1) and the Underlying Chemical and Physical Aerosol Properties." *Journal of Geophysical Research*, 103, no. D13, pp. 16,547-16,563, 1998.

Reheis, M. C., et al. "Dust Deposition in Southern Nevada and California, 1984-1989: Relations to Climate, Source Area, and Lithology." *Journal of Geophysical Research*, 100, no. D5, pp. 8,893-8,918, 1995.

Rogers, C. A., et al. "Evidence of Long-Distance Transport of Mountain Cedar Pollen into Tulsa, Oklahoma." *International Journal of Biometeorology*, 42, pp. 65-72, 1998.

Toy, Edmond, et al. "Fueling Heavy Duty Trucks: Diesel or Natural Gas?" *Risk in Perspective*, 8, no. 1, pp. 1-6, 2000.

U.S. Air Force. "U.S. Air Force Evolved Expendable Launch Vehicle Program: Final Supplemental Environmental Impact Statement." Published at: http://ax.laafb.af. mil/axf/eelv/, 2000.

U.S. Department of Agriculture. "Global Warming's High Carbon Dioxide Levels May Exacerbate Ragweed Allergies." USDA press release. Release No. 0278.00, August 2000.

U.S. Environmental Protection Agency. "National Air Pollution Emission Trends Update, 1900-1998." EPA 454/R-00-002, March 2000. Published at: www.epa.gov/ttn/chief/trends98/emtrnd.html

Westbrook, J. K., et al. "Atmospheric Scales of Biotic Dispersal." *Agricultural and Forest Meteorology*, 97, pp. 263-274, 1999.

6: DUST ON THE WIND HEEDS NO BORDERS

Anderson, Theodore L., et al. "Biological Sulfur, Clouds, and Climate." From *Encyclopedia of Earth System Science*, volume 1. New York: Academic Press, 1992.

Charlson, R. J., et al. "Sulfate Aerosol and Climate Change." *Scientific American*, 270, no. 2, pp. 48-57, 1994.

Darwin, Charles. "An Account of the Fine Dust Which Often Falls on Vessels in the Atlantic Ocean." *Quarterly Journal of the Geological Society of London*, 2, pp. 26-30, 1846.

Delany, A. C., et al. "Airborne Dust Collected at Barbados." *Geochimica et Cosmochimica Acta*, 31, pp. 885-909, 1967.

Derbyshire, E., et al. "Landslides in the Gansu Loess of China." *Catena Supplement*, 20, pp. 119-145, 1991.

Dong, Hai. "Pollution a Culprit in Most Beijing Fogs." *Beijing Wanbao,* January 16, 1999. Published at: www.usembassy-china.org.cn/english/sandt/Bjfog.htm

Florig, H. Keith. "China's Air Pollution Risks." *Environmental Science & Technology,* 31, no. 6, pp. 274A–279A, 1997.

Franzen, Lars G. "The 'Yellow Snow' Episode of Northern Fennoscandia, March 1991—a Case Study of Long-Distance Transport of Soil, Pollen and Stable Organic Compounds." *Atmospheric Environment,* 28, no. 22, pp. 3,587–3,604, 1994.

Fullen, M. A., et al. "Aeolian Processes and Desertification in North Central China." Presented at: Wind Erosion: An International Symposium/Workshop. Manhattan, Kansas, June 3–5, 1997.

Jaffe, D., et al. "Transport of Asian Air Pollution to North America." *Geophysical Research Letters,* 26, pp. 711–714, 1999.

Knipping, E. M., et al. "Experiments and Simulations of Ion-Enhanced Interfacial Chemistry on Aqueous NaCl Aerosols." *Science,* 288, pp. 301–306, April 14, 2000.

Koehler, Birgit G., et al. "An FTIR Study of the Adsorption of SO_2 on n-Hexane Soot from -130° to -40°C." *Journal of Geophysical Research—Atmospheres,* 104, no. D5, pp. 5,507–5,515, 1999.

Landler, Mark. "Choking on China's Air, but Loath to Cry Foul." *New York Times,* February 12, 1999.

Lee, ShanHu, et al. "Lower Tropospheric Ozone Trend Observed in 1989–1997 in Okinawa, Japan." *Geophysical Research Letters,* 25, no. 10, pp. 1,637–1,640, 1998.

Perry, Kevin D., et al. "Long-Range Transport of North African Dust to the Eastern United States." *Journal of Geophysical Research,* 102, no. D10, pp. 11,225–11,238, 1997.

Petit, Charles W. "Weekend Rainouts Could Be Our Own Fault." *U.S. News & World Report,* 125, p. 4, 1998.

Quinn, P. K., et al. "Surface Submicron Aerosol Chemical Composition: What Fraction Is Not Sulfate?" *Journal of Geophysical Research,* 105, no. D5, pp. 6,785–6,806, 2000.

Raloff, J. "Sooty Air Cuts China's Crop Yields." *Science News Online,* December 4, 1999. Published at: www.sciencenews.org/search.asp?target-Sooty+air+cuts+China%27s&goButton=Search&navEvent=Top

Ram, Michael, et al. "Insoluble Particles in Polar Ice: Identification and Measurement of the Insoluble Background Aerosol." *Journal of Geophysical Research,* 21, no. D7, pp. 8,378–8,382, 1994.

Thompson, R. D. *Atmospheric Processes and Systems.* London, New York: Routledge, 1998.

United States Embassy, Beijing, China. "Partial Summary, Comments on 'Can the Environment Wait? Priorities for East Asia.'" Published at: www.usembassy-china.org.cn/english/sandt/bjpollu.htm

———. "PRC Air Pollution: How Bad Is It?" 1998. Published at: www.usembassy-china.org.cn/english/sandt/Airq3wb.htm

U.S. Environmental Protection Agency. *National Air Pollutant Emission Trends, 1900–1998.* March 2000. Published at: www.epa.gov/ttn/chief/trends98/emtrnd.html

Wilson, Richard, and Spengler, John D. *Particles in Our Air: Concentrations and Health Effects.* Boston: Harvard University Press, 1996.

Zhang, X. Y. et al. "Sources, Emission, Regional- and Global-Scale Transport of

Asian Dust." *Conference Proceedings: Dust Aerosols, Loess Soils & Global Change, Washington State University,* 1998.

7: Did Dust Do In the Ice Age?

Ackerman, A. S., et al. "Reduction of Tropical Cloudiness by Soot." *Science,* 288, pp. 1,042–1,047, May 12, 2000.

Anderson, Theodore L., et al. "Biological Sulfur, Clouds and Climate." In *Encyclopedia of Earth System Science.* Nierenberg, William A., ed. Orlando, Fla.: Academic Press, Inc., 1992.

Basile, Isabelle, et al. "Patagonian Origin of Glacial Dust Deposited in East Antarctica (Vostok and Dome C) During Glacial Stages 2, 4 and 6." *Earth and Planetary Science Letters,* 146, pp. 573–589, 1997.

Biscaye, P. E., et al. "Asian Provenance of Glacial Dust (Stage 2) in Greenland Ice Sheet Project 2 Ice Core, Summit, Greenland." *Journal of Geophysical Research,* 102, no. C12, pp. 26,765–26,781, 1997.

Boyd, P. W., et al. "Atmospheric Iron Supply and Enhanced Vertical Carbon Flux in the NE Subarctic Pacific: Is There a Connection?" *Global Biogeochemical Cycles,* 12, no. 3, pp. 429–441, 1998.

Coale, Kenneth H., et al. "A Massive Phytoplankton Bloom Induced by an Ecosystem-Scale Iron Fertilization Experiment in the Equatorial Pacific Ocean." *Nature,* 383, pp. 495–501, October 11, 1996.

Gray, William M., et al. "Weather Modification by Carbon Dust Absorption of Solar Energy." *Journal of Applied Meteorology,* 15, pp. 355–386, April 1976.

Hansen, James, et al. "Global Warming in the Twenty-First Century: An Alternative Scenario." *Proceedings of the National Academy of Science,* 97, no. 18, pp. 9,875–9,880, 2000.

Ledley, T. S., et al. "Potential Effects of Nuclear War Smokefall on Sea Ice." *Climatic Change,* 8, pp. 155–171, 1986.

———. "Sediment-Laden Snow and Sea Ice in the Arctic and Its Impact on Climate." *Climatic Change,* 37, pp. 641–664, 1997.

Li, L.-A., et al. "The Impact of Worldwide Volcanic Activities on Local Precipitation—Taiwan as an Example." *Journal of the Geological Society of China,* 40, pp. 299–311, 1997.

Overpeck, Jonathan, et al. "Possible Role of Dust-Induced Regional Warming in Abrupt Climate Change During the Last Glacial Period." *Nature,* 384, pp. 442–449, December 5, 1996.

Podgorny, I. A., et al. "Aerosol Modulation of Atmospheric and Surface Solar Heating Over the Tropical Indian Ocean." *Tellus,* 52B, pp. 947–958, 2000.

Prospero, Joseph M., et al. "Impact of the North African Drought and El Niño on Mineral Dust in the Barbados Trade Winds." *Nature,* 320, pp. 735–738, April 24, 1986.

Rhodes, Johnathon J. "Mode of Formation of 'Ablation Hollows' Controlled By Dirt Content of Snow." *Journal of Glaciology,* 33, no. 4. pp. 135–139, 1987.

Rosenfeld, D. "TRMM Observed First Direct Evidence of Smoke from Forest Fires Inhibiting Rainfall." *Geophysical Research Letters,* 26, no. 20, pp. 3,105–3,109, 1999.

Rosenfeld, Daniel. "Suppression of Rain and Snow by Urban and Industrial Air Pollution." *Science,* 287, pp. 1,793–1,796, July 14, 2000.

Steen, R. S. "Cryosphere-Atmosphere Interactions in the Global Climate System." Ph.D. Dissertation, Rice University, December 1997.

Taylor, Kendrick. "Rapid Climate Change." *American Scientist,* 87, no. 4, pp. 320–327, 1999.

Tegen, Ina, et al. "The Influence of Climate Forcing of Mineral Aerosols from Disturbed Soils." *Nature,* 380, pp. 419–422, April 4, 1996.

Twohy, C. H., et al. "Light-Absorbing Material Extracted from Cloud Droplets and its Effect on Cloud Albedo." *Journal of Geophysical Research,* 94, no. D6, pp. 8,623–8,631, 1989.

"UW Professor's Climate Change Theory Leads to NASA Mission." University of Washington press release, August 2, 1999.

Warren, S. G. "Impurities in Snow: Effects on Albedo and Snowmelt (Review)." *Annals of Glaciology,* 5, pp. 177–179, 1984.

8: A STEADY *DOWNWARD* RAIN OF DUST

"Bad Decision on Clean Air." *New York Times,* p. A22, May 19, 1999.

Busacca, Alan, ed. *Conference Procedings: Dust Aerosols, Loess Soils, & Global Change, Washington State University.* Publication No. MISC0190, 1998.

Chadwick, O. A., et al. "Changing Sources of Nutrients During Four Million Years of Ecosystem Development." *Nature* 397, pp. 491–497, 1999.

Darwin, Charles. "An Account of the Fine Dust Which Often Falls on Vessels in the Atlantic Ocean." *Quarterly Journal of the Geological Society of London,* 2, pp. 26–30, 1846.

Edworthy, Jason. "Red Snow in the Rockies." *Canadian Alpine Journal,* 61, pp. 71–78, 1978.

Gao, Y., et al. "Relationships Between the Dust Concentrations Over Eastern Asia and the Remote North Pacific." *Journal of Geophysical Research,* 97, pp. 9,867–9,872, 1992.

Health Effects Institute and Aeronomy Laboratory of NOAA. *Report of the PM Measurements Research Workshop, Chapel Hill, North Carolina, 22–23 July, 1998.* Cambridge, Mass.: Health Effects Institute, 1998.

Hefflin, G. J., et al. "Surveillance for Dust Storms and Respiratory Diseases in Washington State, 1991." *Archives of Environmental Health,* 49, no. 3, pp. 170–174, 1994.

Holden, Constance, ed. "Cool DNA." *Science,* 285, p. 327, July 16, 1999.

Hurst, Christon J., ed. *Manual of Environmental Microbiology.* Washington, D.C.: ASM Press, 1997.

Levetin, Estelle. "Aerobiology of Agricultural Pathogens." In *Manual of Environmental Microbiology,* Hurst, Christon J., ed. Washington, D.C.: ASM Press, 1997.

Muhs, Daniel R., et al. "Geochemical Evidence of Saharan Dust Parent Material for Soils Developed on Quaternary Limestones of Caribbean and Western Atlantic Islands." *Quaternary Research,* 33, pp. 157–177, 1990.

NASA. "Magnetite-Producing Bacteria Found in Desert Varnish." NASA Ames press release 97-32, May 1, 1997. Published at: ccf.arc.nasa.gov/dx/basket/stories etc97_32AR.html

Nowicke, Joan W., et al. "Yellow Rain—a Palynological Analysis." *Nature,* 309, pp. 205-207, May 17, 1984.

Perry, Kevin D., et al. "Long-Range Transport of North African Dust to the Eastern United States." *Journal of Geophysical Research,* 102, no. D10, pp. 11,225-11,238, 1997.

Priscu, John C., et al. "Perennial Antarctic Lake Ice: An Oasis of Life in a Polar Desert." *Science,* 280, pp. 2,095-2,098, June 26, 1998.

Prospero, J. M., et al. "Impact of the North African Drought and El Niño on Mineral Dust in the Barbados Trade Winds." *Nature,* 320, pp. 735-738, 1986.

Psenner, Roland. "Living in a Dusty World: Airborne Dust as a Key Factor for Alpine Lakes." *Water, Air and Soil Pollution,* 112, pp. 217-227, 1999.

———, et al. "Life at the Freezing Point." *Science,* 280, pp. 2,073-2,074, June 26, 1998.

Reheis, M. C., et al. "Dust Deposition in Southern Nevada and California, 1984-1989: Relations to Climate, Source Area, and Lithology." *Journal of Geophysical Research,* 100, no. D5, pp. 8,893-8,918, 1995.

Reuther, Christopher G. "Winds of Change: Reducing Transboundary Air Pollutants." *Environmental Health Perspectives,* 108, no. 4, pp. A170-175, 2000.

Schlesinger, Richard B. "Properties of Ambient PM Responsible for Human Health Effects: Coherence Between Epidemiology and Toxicology." *Inhalation Toxicology,* 12 (Suppl. 1), pp. 23-25, 2000.

Shinn, Eugene A., et al. "African Dust and the Demise of Caribbean Coral Reefs." *Geophysical Research Letters,* 27, no. 19, pp. 3,029-3,033, 2000.

———. "139 Bacteria and Fungi Isolated from African Dust." Personal communication, February 2001.

Silver, Mary W., et al. "Ciliated Protozoa Associated with Oceanic Sinking Detritus." *Nature,* 309, pp. 246-248, May 17, 1984.

———. "The 'Particle' Flux: Origins and Biological Components." *Progress in Oceanography,* 26, pp. 75-113, 1991.

Smith, G. T., et al. "Caribbean Sea Fan Mortalities." *Nature,* 383, pp. 487, 1996.

Stone, Richard. "Lake Vostok Probe Faces Delays." *Science,* 286, pp. 36-37, October 1, 1999.

Swap, R., et al. "Saharan Dust in the Amazon Basin." *Tellus,* 44B, pp. 133-149, 1992.

Toy, Edmond, et al. "Fueling Heavy Duty Trucks: Diesel or Natural Gas?" *Risk in Perspective,* 8, no. 1, pp. 1-6, 1999.

U.S. Centers for Disease Control. "National Vital Statistics Report." 47, 1998.

U.S. Environmental Protection Agency. *Deposition of Air Pollutants to the Great Waters: Second Report to Congress.* USEPA, Office of Air Quality. Research Triangle Park, June 1997. EPA-453/R-97-011.

———. "Nonattainment Designations for PM-10 as of August 1999." Published at: www.epa.bgov/airs/rvnonpm1.gif

Weiss, P., et al. "Impact, Metabolism and Toxicology of Organic Xenobiotics in Plants: A summary of the 4th IMTOX-Workshop contents." Published at: www.ubavie.gv.at/publikationen/tagungs/CP24s.HTM

Wright, Robert J., et al. *Agricultural Uses of Municipal Animal and Industrial Byproducts.*

Washington, D.C.: USDA. 1998. Published at: www.ars.usda.gov/is/np/agbyprod ucts/agbyintro.htm

Young, R. W., et al. "Atmospheric Iron Inputs and Primary Productivity: Phytoplankton Responses in the North Pacific." *Global Biochemical Cycles,* 5, no. 2, pp. 119-134, 1991.

9: A Few Unsavory Characters from the Neighborhood

Ataman, G. "The Zeolitic Tuffs of Cappadocia and Their Probable Association with Certain Types of Lung Cancer and Pleural Mesothelioma." *Comptes Rendus de l'Académie de Science* (Paris), 287, pp. 207-210, 1978.

Baris, Y. I., et al. "An Outbreak of Pleural Mesothelioma and Chronic Fibrosing Pleurisy in the Village of Karain/Ürgüp in Anatolia." *Thorax,* 33, pp. 181-192, 1978.

Baum, Gerald L., et al. *Textbook of Pulmonary Diseases,* sixth edition. Philadelphia: Lippincott-Raven, 1997.

Beckett, William, et al. "Adverse Effects of Crystalline Silica Exposure." (Official Statement of the American Thoracic Society.) *American Journal of Critical Care Medicine,* 155, pp. 761-768, 1997.

Brambilla, Christian, et al. "Comparative Pathology of Silicate Pneumoconiosis." *American Journal of Pathology,* 96, no. 1, pp. 149-169, 1979.

Christensen, L. T., et al. "Pigeon Breeders' Disease—a Prevalence Study and Review." *Clinical Allergy,* 5, no. 4, pp. 417-430, 1975.

Cockburn, Aidan, et al. "Autopsy of an Egyptian Mummy." *Science,* 187, no. 4, 182, pp. 1,155-1,160, 1975.

Dong, Depu, et al. "Lung Cancer Among Workers Exposed to Silica Dust in Chinese Refractory Plants." *Scandinavian Journal of Work and Environmental Health,* 21 (Suppl. 2), pp. 69-72, 1995.

Englehardt, James, principal investigator. "Solid Waste Management Health and Safety Risks: Epidemiology and Assessment to Support Risk Reduction." Florida Center for Solid and Hazardous Waste Management. Gainesville, 1999.

Grobbelaar, J. P. "Hut Lung: A Domestically Acquired Pneumoconiosis of Mixed Aetiology in Rural Women." *Thorax,* 46, pp. 334-341, 1991.

Harris, Gardiner, et al. "Dust, Death & Deception: Why Black Lung Hasn't Been Wiped Out." *Courier-Journal,* April 19-26, 1998. Published at: www.courier-journal. com/dust/index.html

Hirsch, Menachem, et al. "Simple Siliceous Pneumoconiosis of Bedouin Females in the Negev Desert." *Clinical Radiology,* 25, pp. 507-510, 1974.

Homes, M. J., et al. "Viability of Bioaerosols Produced from a Swine Facility." Proceedings, International Conference on Air Pollution from Agricultural Operations, Kansas City, Mo., pp. 127-132, February 7-9, 1996.

Houba, Remko, et al. "Occupational Respiratory Allergy in Bakery Workers: A Review of the Literature. *American Journal of Industrial Medicine,* 34, no. 6, pp. 529-546, 1998.

Hubbard, Richard, et al. "Occupational Exposure to Metal or Wood Dust and Aetiology of Cryptogenic Fibrosing Alveolitis." *Lancet,* 347, no. 8997, pp. 284–289, 1996.

Korenyi-Both, A. L., et al. "Al Eskan Disease: Desert Storm Pneumonitis." *Military Medicine,* 157, no. 9, pp. 452–462, 1992.

———. "The Role of the Sand in Chemical Warfare Agent Exposure among Persian Gulf War Veterans: Al Eskan Disease and 'Dirty Dust'." *Military Medicine,* 165, no. 5, pp. 321–336, 2000.

Leigh, J. Paul, et al. "Occupational Injury and Illness in the United States." *Archives of Internal Medicine,* 157, pp. 1,557–1,568, 1997.

Lemieux, Paul M., et al. "Emissions of Polychlorinated Dibenzo-p-dioxins and Polychlorinated Dibenzofurans from the Open Burning of Household Waste in Barrels." *Environmental Science and Technology,* 34, no. 3, pp. 377–384, 2000.

Lilis, Ruth. "Fibrous Zeolites and Endemic Mesothelioma in Cappadocia, Turkey." *Journal of Occupational Medicine,* 23, no. 8, pp. 548–550, 1981.

Linch, Kenneth D., et al. "Surveillance of Respirable Crystalline Silica Dust Using OSHA Compliance Data (1979-1995)." *American Journal of Industrial Medicine,* 34, pp. 547–558, 1998.

Marsden, William, trans. and ed. Wright, Thomas, re-ed. *The Travels of Marco Polo the Venetian.* New York: Doubleday & Company, Inc., 1948.

Miguel, Ann G., et al. "Allergens in Paved Road Dust and Airborne Particles." *Environmental Science & Technology,* 33, no. 23, pp. 4,159–4,168, 1999.

Morgan, W. Keith, et al. *Occupational Lung Diseases.* Philadelphia. W. B. Saunders, 1984.

Mumpton, Frederick A. "Report of Reconnaissance Study of the Association of Zeolites with Mesothelioma Cancer Occurrences in Central Turkey." Brockport, N.Y. Department of Earth Sciences, State University College, 1979.

National Institute for Occupational Safety and Health. "Preventing Asthma in Animal Handlers." Department of Health and Human Services (NIOSH) Publication No. 97-116, 1998.

———. "Request for Assistance in Preventing Organic Dust Toxic Syndrome." Department of Health and Human Services (NIOSH) Publication No. 94-102, 1994.

Norboo, T., et al. "Silicosis in a Himalayan Village Population: Role of Environmental Dust." *Thorax,* 46, pp. 341–343, 1991.

OSHA. "Cotton Dust." OSHA Fact Sheet: 95-23, 1995.

———. "Grain Dust (Oat, Wheat and Barley)." Comments from the June 19, 1988, Final Rule on Air Contaminants Project extracted from 54FR2324 et. seq. Published at: www.cdc.gov/niosh/pel88/graindst.html

Pless-Mulloli, T., et al. "Living Near Opencast Coal Mining Sites and Children's Respiratory Health." *Occupational & Environmental Medicine,* 57, no. 3, pp. 145–151, 2000.

Reuters News Service. "Ex-U.S. Army Doctor Says Uranium Shells Harmed Vets." Reuters, September 3, 2000.

Rodrigo, M. J., et al. "Detection of Specific Antibodies to Pigeon Serum and Bloom Antigens by Enzyme Linked Immunosorbent Assay in Pigeon Breeder's Disease." *Occupational & Environmental Medicine,* 57, no. 3, pp. 159–164, 2000.

Rohl, H. N., et al. "Endemic Pleural Disease Associated with Exposure to Mixed Fibrous Dust in Turkey." *Science,* 216, pp. 518–520, 1982.

Schneider, Andrew. "Asbestos-Containing Gardening Product Still Being Sold in Seattle Area." *Seattle Post-Intelligencer,* March 31, 2000. Published at: www.seattle-pi.com/uncivilaction/

——. "Uncivil Action." *Seattle Post-Intelligencer,* November 18–19, 1999. Published at: www.seattle-pi.com/uncivilaction/

Schneider, Eileen, et al. "A Coccidioidomycosis Outbreak Following the Northridge, Calif., Earthquake." *Journal of the American Medical Association,* 277, no. 11, pp. 905–909, 1997.

Schwartz, L. W., et al. "Silicate Pneumoconiosis and Pulmonary Fibrosis in Horses from the Monterey-Carmel Peninsula." *Chest,* 80 (suppl.), pp. 82S–85S, 1981.

Sebastien, P., et al. "Zeolite Bodies in Human Lungs from Turkey." *Laboratory Investigation,* 44, no. 5, pp. 420–425, 1981.

Sherwin, R. P., et al. "Silicate Pneumoconiosis of Farm Workers." *Laboratory Investigation,* 40, no. 5, pp. 576–581, 1979.

Simpson, J. C., et al. "Comparative Personal Exposures to Organic Dusts and Endotoxin." *Annals of Occupational Hygiene,* 43, no. 2, pp. 107–115, 1999.

U.S. Centers for Disease Control. "Hantavirus Pulmonary Syndrome—Panama, 1999–2000." *Morbidity and Mortality Weekly Report,* 49, no. 10, pp. 205–207, 2000.

——. *Work-Related Lung Disease Surveillance Report 1999.* Washington, D.C., 1999.

——. *Occupational Exposure to Respirable Coal Mine Dust.* Washington, D.C., 1996.

——. "Respiratory Illness Associated with Inhalation of Mushroom Spores—Wisconsin, 1994." *Morbidity and Mortality Weekly Report,* 43, no. 29, pp. 525–526, 1994.

U.S. Department of Labor. "Labor Secretary Hits Fraud in Coal Mine Health Sampling Program." USDOL Office of Information. 91-151, 1991.

U.S. Environmental Protection Agency. "National Air Pollution Emission Trends Update, 1900–1998." EPA 454/R-00-002, March 2000. Published at: www.epa.gov/ttn/chief/trends98/emtrnd.html

——. "The 1998 Toxic Release Inventory." EPA-745-R-00-002. Published at: www.epa.gov/tri/tri98/index.htm

Zenz, Carl, et al. *Occupational Medicine,* third edition. St. Louis: Mosby, 1994.

Zhong, Yuna, et al. "Potential Years of Life Lost and Work Tenure Lost When Silicosis Is Compared with Other Pneumoconioses." *Scandinavian Journal of Work and Environmental Health,* 21 (Suppl. 2), pp. 91–94, 1995.

10: MICROSCOPIC MONSTERS AND OTHER INDOOR DEVILS

Abt, Eileen, et al. "Characterization of Indoor Particle Sources: A Study Conducted in the Metropolitan Boston Area." *Environmental Health Perspectives,* 108, no. 1, pp. 35–44, 2000.

Anderson, Rosalind C., et al. "Toxic Effects of Air Freshener Emissions." *Archives of Environmental Health,* 52, pp. 433–441, 1997.

Arlian, L. G., et al. "Population Dynamics of the House Dust Mites *Dermatophagoides farinae, D. pteronyssinus and Euroglyphus maynei* (Acari: Pyroglyphidae) at Specific Relative Humidities." *Journal of Medical Entomology,* 35, no. 1, pp. 46–53, 1998.

Arlian, Larry G., et al. "Prevalence of Dust Mites in the Homes of People with Asthma Living in Eight Different Geographic Areas of the United States." *Journal of Allergy and Clinical Immunology,* 90, no. 3, pp. 292–300, 1992.

Ball, Thomas M., et al. "Siblings, Day-Care Attendance, and the Risk of Asthma and Wheezing During Childhood." *New England Journal of Medicine,* 343, no. 8, pp. 538–543, 2000.

Bernstein, Nina. "38% Asthma Rate Found in Homeless Children." *New York Times,* pp. B1, B13, May 5, 1999.

Bodner, C., et al. "Childhood Exposure to Infection and Risk of Adult Onset Wheeze and Atopy." *Thorax,* 55, pp. 28–32, 2000.

———. "Family Size, Childhood Infections and Atopic Diseases. The Aberdeen WHEASE Group." *Thorax,* 53, pp. 383–387, 1998.

Burge, Harriet A., ed. *Bioaerosols (Indoor Air Research).* Boca Raton: Lewis Publishers, 1995.

Chapman, M. D. "Environmental Allergen Monitoring and Control." *Allergy,* 53, pp. 48–53, 1998.

Christie, G. L., et al. "Is the Increase in Asthma Prevalence Occurring in Children without a Family History of Atopy?" *Scottish Medical Journal,* 43, no. 6, pp. 180–182, 1998.

Cralley, Lester V., et al. *Health and Safety Beyond the Workplace.* New York: John Wiley & Sons, Inc., 1990.

Cramer, Daniel W., et al. "Genital Talc Exposure and Risk of Ovarian Cancer." *International Journal of Cancer,* 81, pp. 351–356, 1999.

Crater, Scott E., et al. "Searching for the Cause of the Increase in Asthma." *Current Opinion in Pediatrics,* 10, pp. 594–599, 1998.

Cizdziel, James V., et al. "Attics as Archives for House Infiltrating Pollutants: Trace Elements and Pesticides in Attic Dust and Soil from Southern Nevada and Utah." *Microchemical Journal,* 64, pp. 85–92, 2000.

Duff, Angela L., et al. "Risk Factors for Acute Wheezing in Infants and Children: Viruses, Passive Smoke, and IgE Antibodies to Inhalant Allergens." *Pediatrics,* 92, no. 4, pp. 535–540, 1993.

Egan, A. J., et al. "Munchausen Syndrome Presenting as Pulmonary Talcosis." *Archives of Pathology and Laboratory Medicine,* 123, pp. 736–738, 1999.

Erb, Klaus J. "Atopic disorders: A Default Pathway in the Absence of Infection?" *Immunology Today,* 20, pp. 317–322, 1999.

Ernst, Pierre, et al. "Relative Scarcity of Asthma and Atopy Among Rural Adolescents Raised on a Farm." *American Journal of Respiratory and Critical Care Medicine,* 161, no. 5, pp. 1,563–1,566, 2000.

Farooqi, Sadaf I., et al. "Early Childhood Infection and Atopic Disorder." *Thorax,* 53, pp. 927–932, 1998.

Felton, James. "Mutagens: The Role of Cooked Food in Genetic Changes." *Science & Technology Review,* pp. 6–24, July 1995.

Finkelman, Robert B., et al. "Health Impacts of Domestic Coal Use in China." *Proceedings of the National Academy of Sciences,* 96, pp. 3,427–3,431, 1999.

Florig, Keith. "China's Air Pollution Risks." *Environmental Science & Technology*, 31, no. 6, pp. 274A–279A 1997.

Gereda, J. E., et al. "Relation Between House-Dust Endotoxin Exposure, Type I T-cell Development, and Allergen Sensitisation in Infants at High Risk of Asthma." *Lancet*, 355, no. 9,216, pp. 1,680–1,683, 2000.

Hagmann, Michael. "A Mold's Toxic Legacy Revisited." *Science*, 288, pp. 243–244, 2000.

Holgate, S. T. "Asthma and Allergy—Disorders of Civilization?" *QJM*, 91, pp. 171–184, 1998.

Hopkin, J. M. "Atopy, Asthma, and the Mycobacteria." *Thorax*, 55, pp. 454–458, 2000.

Knize, Mark G., et al. "The Characterization of the Mutagenic Activity of Soil." *Mutation Research*, 192, pp. 23–30 1987.

Lewis, S. A. "Infections in Asthma and Allergy." *Thorax*, 53, pp. 911–912, 1998.

Lioy, Paul J., et al. "Air Pollution." Environmental and Occupational Health Sciences Institute Web site: http://snowfall.envsci.rutgers.edu/~kkeating/101_html/101syllabus_html/lect21-AirPollution_html/ted01_html/

Matricardi, Paolo M., et al. "Exposure to Foodborne and Orofecal Microbes versus Airborne Viruses in Relation to Atopy and Allergic Asthma: Epidemiological Study." *British Medical Journal* 320, pp. 412–417, 2000.

Motomatsu, Kenichi, et al. "Two Infant Deaths After Inhaling Baby Powder." *Chest*, 75, pp. 448–450, 1979.

Nilsson, L., et al. "A Randomized Controlled Trial of the Effect of Pertussis Vaccines on Atopic Disease." *Archives of Pediatric and Adolescent Medicine*, 152, pp. 734–738, 1998.

Nishioka, M., et al. "Measuring Transport of Lawn-Applied Herbicide Acids from Turf to Home: Correlation of Dislodgeable 2,4-D Turf Residues with Carpet Dust and Carpet Surface Residues." *Environmental Science & Technology*, 30, pp. 3,313–3,320, 1996.

Nishioka, M. G., et al. "Measuring Transport of Lawn-Applied 2,4-D and Subsequent Indoor Exposures of Residents." Abstract of meeting paper published at: www.riskworld.com/Abstract/1996/SRAam96/ab6aa159.htm

Ott, Wayne R., et al. "Everyday Exposure to Toxic Pollutants." *Scientific American*, pp. 86–93, February 1998.

Ozkaynak, J. Xue, et al. "The Particle Team (PTEAM) Study: Analysis of the Data." USEPA Project Summary.EPA/600/SR-95/098, April 1997.

Pew Environmental Health Commission. "Attack Asthma: Why America Needs a Public Health Defense System to Battle Environmental Threats," May 16, 2000. Published at: pewenvirohealth.jhsph.edu/html/reports/PEHCAsthmaReport.pdf

Platts-Mills, Thomas A. E., et al. "Indoor Allergens and Asthma: Report of the Third International Workshop." *Allergy and Clinical Immunology*, 100, no. 6, pp. S2–S24, 1997.

Public Citizen. "Letter to Ann Brown, Chairperson, U.S. Consumer Product Safety Commission." (Re: lead-core candle wicks.) February 24, 2000. Published at: www.citizen.org/hrg/PUBLICATIONS/1510.htm#Supplemental letter.

Reed, K., et al. "Quantification of Children's Hand and Mouthing Activity." *Journal of Exposure Analysis and Environmental Epidemiology*, 9, no. 5, pp. 513–520, 1999.

Roberts, J. W., et al. "Reducing Dust, Lead, Dust Mites, Bacteria, and Fungi in Carpets by Vacuuming." *Archives of Environmental Contamination and Toxicology,* 36, pp. 477–484, 1999.

Robin, L. F., et al. "Wood-Burning Stoves and Lower Respiratory Illnesses in Navajo Children." *Pediatric Infectious Disease,* 15, no. 10, pp. 859–865, 1998.

Rogge, Wolfgang F., et al. "Sources of Fine Organic Aerosol. Part 6. Cigarette Smoke in the Urban Atmosphere." *Environmental Science & Technology,* 28, pp. 1,375–1,388, 1994.

Sigurs, Nele, et al. "Respiratory Syncytial Virus Bronchiolitis in Infancy Is an Important Risk Factor for Asthma and Allergy at Age 7." *American Journal of Respiratory and Critical Care Medicine,* 11, no. 5, pp. 1,501–1,507, 2000.

Tariq, S. M., et al. "The Prevalence of and Risk Factors for Atopy in Early Childhood: A Whole Population Birth Cohort Study." *Journal of Allergy and Clinical Immunology,* 101, no. 5, pp. 587–593, 1998.

Thiebaud, Herve P., et al. "Mutagenicity and Chemical Analysis of Fumes from Cooking Meat." *Journal of Agricultural and Food Chemistry,* 42, no. 7, pp. 1502–1510, 1994.

U.S. Centers for Disease Control. "Acute Pulmonary Hemorrhage/Hemosiderosis Among Infants." *Morbidity and Mortality Weekly Report,* 43, no. 48, pp. 881–883, 1994.

U.S. Consumer Product Safety Commission. "CPSC Finds Lead Poisoning Hazard for Young Children in Imported Vinyl Miniblinds." Press release #96-150. 1996.

———. U.S. Environmental Protection Agency. "Use and Care of Home Humidifiers." 1991. Published at www.epa.gov/iaq/pubs/humidif.html

U.S. Environmental Protection Agency. *Respiratory Health Effects of Passive Smoking.* Office of Research and Development, Office of Air and Radiation. EPA-43-F-93-003. 1993.

U.S. Food and Drug Administration. *The Food Defect Action Levels Handbook.* Center for Food Safety and Applied Nutrition. Revised May 1998. Published at: http://vm.cfsan.fda.gov/~dms/dalbook.html

U.S. General Accounting Office. "Indoor Pollution: Status of Federal Research Activities." GAO/RCED-99-254. August, 1999.

van Bronswijk, Johanna E. M. H. *House Dust Biology: For Allergists, Acarologists and Mycologists.* Holland. Published by the author, 1981. Distributed by the author: j.e.m.h.v.bronswijk@allergo.nl

Wallace, Lance. "Correlations of Personal Exposure to Particles with Outdoor Air Measurements: A Review of Recent Studies." *Aerosol Science and Technology,* 32, no. 1, pp. 15–26, 2000.

———. "Real-Time Monitoring of Particles, PAH, and CO in an Occupied Townhouse." *Applied Occupational and Environmental Hygiene,* 15, no. 1, pp. 39–47, 2000.

Wouters, I. M., et al. "Increased Levels of Markers of Microbial Exposure in Homes with Indoor Storage of Organic Household Waste." *Applied Environmental Microbiology,* 66, no. 2, pp. 627–631, 2000.

11: DUST TO DUST

Adams, Fred, et al. *The Five Ages of the Universe: Inside the Physics of Eternity.* New York: The Free Press, 1999.

Bay Area Air Quality Management District. *Permit Handbook.* Published at: www.baaqmd.gov/permit/handbook/s11c05ev.htm

Cremation Association of North America. "History of Cremation." Published at: www.cremationassociation.org/html/history.html

———. "Emissions Tests Provide Positive Results for Cremation Industry." 1999. Published at: www.cremationassociation.org/html/environment.html

———. "1998 Cremation Data and Projections to the Year 2010." 1999. Published at: www.cremationassociation.org/html/statistics.html

Irion, Paul E. *Cremation.* Philadelphia: Fortress Press, 1968.

Iserson, Kenneth V. *Death to Dust,* second edition. Tucson, Ariz. Galen Press, Ltd., 2000.

National Park Service. "Southwest Region Parks: Protecting Cultural Heritage." NPS Southwest Region, Santa Fe. U.S. Government Printing Office, pp. 837–845, 1992.

Willson, L. A., et al. "Mass Loss at the Tip of the AGB." Published at: www.public.iastate.edu/~lwillson/homepage.html

———. "Miras, Mass-Loss, and the Ultimate Fate of the Earth." Comments to AAAS, 2000. Published at: www.public.iastate.edu/~lwillson/homepage.html

INDEX